BACTERIAL CELL CULTURE

ESSENTIAL DATA SERIES

Series Editors

D. Rickwood
Department of Biological and Chemical Sciences,
University of Essex, Wivenhoe Park, Colchester, UK

B.D. Hames
Department of Biochemistry and Molecular Biology,
University of Leeds, Leeds, UK

Published titles

Centrifugation
Gel Electrophoresis
Light Microscopy
Vectors
Human Cytogenetics
Animal Cells: culture and media
Cell and Molecular Biology
PCR
Nucleic Acid Hybridization

Immunoassays
Transcription Factors
Enzymes in Molecular Biology
Neuropeptides
Liquid Chromatography
Bacterial Cell Culture

Forthcoming titles

Cytokines
Growth Factors
Enzyme Assays

See final pages for full list of titles and order form

BACTERIAL CELL CULTURE
ESSENTIAL DATA

A.S. Ball

Department of Biological and Chemical Sciences, University of Essex, Colchester, UK

JOHN WILEY & SONS

Chichester · New York · Weinheim · Brisbane · Singapore · Toronto

Published in association with BIOS Scientific Publishers Limited

©1997 John Wiley & Sons Ltd, Baffins Lane, Chichester, West Sussex PO19 1UD, UK, tel (01243) 779? association with BIOS Scientific Publishers Ltd, 9 Newtec Place, Magdalen Road, Oxford OX4 1RE, UK.

British Library Cataloguing in Publication Data
A catalogue record for this book is available from the British Library.

ISBN 0471 96973 7

Library of Congress Cataloging in Publicat
Ball, A. S.
Bacterial cell culture: essential data/A.S. Ball
    p. cm.——(Essential data series)
Published in association with BIOS Scientific Publisher
    Includes bibliographical references and index.
    ISBN 0-471-96973-7 (alk. paper)
    1. Bacteriology——Laboratory manuals. 2. Ba
manuals. 3. Cell culture——Laboratory manuals. I.
    [DNLM: 1. Cell Culture——Laboratory manuals.
3. Bacteriological Techniques. QW 25 B187b 1997].
QR63.B25 1997
571.6'.38293——dc21
DNLM/DLC
for Library of Congress

Typeset by Marksbury Multimedia Ltd, Midsomer Norton, Bath, UK
Printed and bound in UK by Page Bros (Norwich) Ltd, UK

CONTENTS

Contents

ABBREVIATIONS

API	analytical profile index
CFU	colony forming units
dNTP	deoxynucleotidetriphosphate
ENCISE	enterobacteriacae numerical coding and identification system for enterotube
FITC	fluorescein isothiocyanate
HPLC	high performance liquid chromatography
MPN	most probable number
TAE	Tris-acetate-EDTA
T_m	melting temperature
TMPD	tetramethylparaphenylene-diamine
UV	ultraviolet
VP (test)	Vosges–Proskauer (test)

PREFACE

This book aims to introduce a range of important bacterial techniques to students who are already familiar with the basic principles of biology and chemistry and who are about to start laboratory work with bacterial cell cultures. The emphasis is on basic laboratory techniques commonly required during the growth, isolation and characterization of bacteria. The subject area is vast and no attempt has been made to cover it in its entirety. Instead I have focused on a few, select areas which I consider to be the most widely used techniques in bacterial cell culture. The information is presented as a series of tables, arranged into seven chapters.

Chapter 1 gives details of the basic requirements needed for the successful cultivation of bacteria, along with details of the types of media commonly used for cultivating heterotrophic, autotrophic and anaerobic bacteria. Chapter 2 details a range of methods used for the isolation of various bacteria from different environments. Chapter 3 describes the different techniques used to aid the microscopic observation of bacteria. This includes a series of stains commonly used to observe selected structures in bacteria. In Chapter 4 the theory of bacterial growth is reviewed and a description of the techniques used to follow both total and viable growth is included. Chapter 5 describes a range of biochemical and cellular techniques used in the identification, characterization and classification of bacteria. In Chapter 6, I aimed to cover the basic requirements for PCR analysis and interpretation. This chapter covers the isolation of bacterial DNA and the subsequent PCR amplification, purification and analysis of target DNA sequences. Chapter 7 provides details of the media used in the cultivation of actinomycetes and the subsequent tentative identification of the actinomycete to the species level. In the final chapter information of manufacturers supplying reagents and equipment is provided. Finally, references for further reading are presented.

A.S. Ball

Chapter 1 CULTIVATION OF BACTERIA

1 Introduction

When bacteria are observed in their natural habitat they can rarely be identified without further cultivation. It is usual to obtain pure cultures of these bacteria by growing them in the laboratory only then can they be identified through a series of cultural, morphological and biochemical properties. However, the metabolic diversity of bacteria allows the selection of specific organisms with specific properties. Generally all isolation and cultivation media require water, an energy source, usually a carbon compound, and a range of other essential elements such as N, P, K, S, Ca and Fe. Some general hints when preparing media for bacterial cultivation are as follows.

1. Use the same source of chemicals for any set of experiments.

2. Dissolve chemicals in distilled or demineralized water. Tap water may contain toxic compounds.

3. Adjust the pH of the media to a suitable value for growth of the bacteria. To determine changes in pH during the growth of bacteria indicators may be added to the medium.

4. Buffers are often required to maintain the pH of the media at a suitable level during growth of the bacteria. Table 1 gives the composition of a range of buffers commonly used in media.

2 Cultures

All bacteria were once considered as variant forms of plants, and some terminology applied to them is derived from that used for plants. The process of growing bacteria is referred

Cultivation of Bacteria

to as culturing, the organisms grown within a confined area being designated a culture. *Table 2* shows the common forms of cultures used in bacteriology.

3 Culture incubations

Following appropriate labeling, cultures should be incubated at the optimum temperature for growth or the activity that is being studied. Generally pathogenic and commensal bacteria grow best at 37°C, while isolates from the environment generally grow best at lower temperatures (10–30°C). However, the physiological diversity of bacteria means that the temperature range 0–100°C, and beyond, can be used to isolate bacteria.

4 Types of media used for cultivating bacteria

A culture medium is any substance, or mixture of substances, used to grow one or more kinds of bacteria in the laboratory. Bacteriological media are frequently classi-

of heterotrophic bacteria using either an organic or an inorganic nitrogen source are shown in *Tables 3* and *4* respectively.

4.2 Autotrophic bacteria

Autotrophs are able to use carbon dioxide as the sole source of carbon for growth. Similarly these organisms can use inorganic sources of nitrogen, phosphate and sulfur. Autotrophic bacteria are therefore cultivated in a mineral salts medium containing no organic compounds.

There are two major types of autotrophic bacteria.

1. Photosynthetic autotrophs obtain energy from light and a source of reducer for the incorporation of either carbon dioxide (blue–green bacteria), hydrogen sulfide (purple and green sulfur bacteria) or organic acids (purple nonsulfur bacteria).

2. Chemosynthetic autotrophs oxidize exogenous inorganic substances to provide both energy and a source of

fied as complex or defined according to their chemical composition. These terms refer to our knowledge of their chemical composition. A complex medium is one that contains such ill-defined substances as yeast or beef extract, while a defined medium can have 10–20 different chemicals present, but the identity of each is known. The sources of energy used by bacteria are diverse and the metabolic range of bacteria and the growth media used for their cultivation are described in the following section.

4.1 Heterotrophic bacteria

All heterotrophic bacteria require an organic carbon source. The range of different media available for the isolation of bacteria reflects the metabolic diversity of these bacteria. Some media do not require any nitrogen source, while others require an organic nitrogen source, while yet others require an inorganic source. Generally, bacteria which require few substrates for growth, including autotrophs, can be grown in simple defined media, while bacteria requiring a complex array of substrates, such as many pathogens, are grown on complex natural media. A range of media used for cultures

reductant, for example nitrate and nitrite (nitrifying bacteria), sulfur and sulfide (sulfur-oxidizing bacteria).

As a consequence of the requirements of these two groups, the media used to cultivate autotrophs are variable. *Table 5* shows a range of media used for the isolation of nitrogen-fixing bacteria. *Table 6* shows the composition of a range of media used for the isolation of photosynthetic bacteria. The media commonly used for the isolation of chemosynthetic bacteria are shown in *Table 7*.

4.3 Anaerobic bacteria

Most bacteria are facultative aerobes, that is they grow under aerobic or anaerobic conditions. Some bacteria only grow in the presence of oxygen; they are obligate aerobes. A few bacteria grow best in low oxygen tensions; they are microaerophilic bacteria. Anaerobes occur in a number of families of bacteria. They include several spore-forming pathogens (e.g. *Clostridium* spp.).

Media used to grow anaerobic bacteria must be heated to

100°C for 10 min to drive off dissolved oxygen, and cooled quickly immediately before use. A number of methods have been devised to remove oxygen from cultures (*Table 8*). For the cultivation of anaerobic bacteria the culture medium must also be reduced. The methods used to achieve this are shown in *Table 9*.

Table 1. Composition of three buffers with a combined pH range of 3.0–10.0; citrate buffer, phosphate buffer and carbonate buffer

Buffer	Comments		
Citrate buffer	Make up 0.1 M of stock solutions. (1) $C_6H_8O_7$ (19.2 g in 1 l) (2) $C_6H_5Na_3O_7.2H_2O$ (29.4 g in 1 l). To make the buffer the required pH, add together the following amounts (ml) of stock solutions 1 and 2 and dilute to 100 ml with distilled water:		
	Required pH	Stock solution 1	Stock solution 2
	3.0	46.5	3.5
	3.2	43.7	6.3
	3.4	40.0	10.0
	3.6	37.0	13.0
	3.8	35.0	15.0
	4.0	33.0	17.0
	4.2	31.5	18.5
	4.4	28.0	22.0
	4.6	25.5	24.5
	4.8	23.0	27.0
	5.0	20.5	29.5
	5.2	18.0	32.0
	5.4	16.0	34.0
	5.6	13.7	36.3
	5.8	11.8	38.2
	6.0	9.5	40.5

Continued

5

Cultivation of Bacteria

Table 1. Composition of three buffers with a combined pH range of 3.0–10.0; citrate buffer, phosphate buffer and carbonate buffer, *continued*

Buffer	Comments
Phosphate buffer	Make up 0.2 M of stock solutions. (1) NaH_2PO_4 (24.0 g in 1 l) (2) Na_2HPO_4 (28.4 g in 1 l). To make the buffer the required pH, add together the following amounts (ml) of stock solutions 1 and 2 and dilute to 200 ml with distilled water:

Required pH	Stock solution 1	Stock solution 2
6.0	87.7	12.3
6.2	81.5	18.5
6.4	73.5	26.4
6.6	62.5	37.5
6.8	51.0	49.0
7.0	39.0	61.0
7.2	28.0	72.0
7.4	19.0	81.0
7.6	13.0	87.0
7.8	8.5	91.5
8.0	5.3	94.7

Buffer	Comments
Carbonate buffer	Make up 0.2 M of stock solutions. (1) Na_2CO_3 (21.2 g in 1 l) (2) $NaHCO_3$ (16.8 g in 1 l). To make the buffer the required pH, add together the following amounts (ml) of stock solutions 1 and 2 and dilute to 100 ml with distilled water:

Required pH	Stock solution 1	Stock solution 2
9.2	4.0	46.0
9.4	9.5	40.5

9.6	16.0	34.0
9.8	22.0	28.0
10.0	27.5	22.5
10.2	33.0	17.0
10.4	38.5	11.5
10.6	42.5	7.5

Table 2. Types of cultures commonly used for the growth of bacteria

Type of culture	Description
Liquid (broth)	Liquid media dispensed into flasks or tubes and plugged with foam plugs or plastic caps. Nonabsorbent cotton wool and screw-cap bottles may also be used
Solid (agar)	Medium can be solidified with agar (usually 1.0–2.0%, see manufacturer's recommendation). Agar, a sulfate ester of a galactan extracted from seaweed, is generally inert. Prior to being dispensed into bottles all agar media must be steamed (or microwaved) until dissolved. The agar will not set until a temperature of 42°C is reached
	Solid media may be used as follows:
	Slope cultures: tubes containing small amounts of medium (5–10 ml) are heated to melt the agar. The agar is allowed to set in a sloping position, increasing the surface area of the medium for the growth of bacteria
	Stab cultures: tubes containing 10–20 ml of solid medium are heated and the medium allowed to cool to 45°C. While liquid the medium is inoculated, mixed and allowed to set

Continued

Cultivation of Bacteria

Table 2. Types of cultures commonly used for the growth of bacteria, *continued*

Type of culture	Description
	Plate cultures: approximately 20–25 ml of solid medium are required for each petri dish (9 cm diameter). The medium is melted but must then be cooled to 45°C before being poured into the dish. The medium can be inoculated using either the poured plate or streak plate methods (see Chapter 2)
Solid (gel)	Occasionally it may be necessary to exclude all organic material (including agar) from media. Silica gel is used in place of agar to solidify the medium. Into distilled water (100 ml) dissolve 13.4 g $Na_2SiO_3.9H_2O$. Adjust the pH to 10.0 by addition of Bio-Rad Laboratories cation exchange resin AG50 WX-8 under constant mixing. After 30 min re-adjust the pH to 10.0 using more resin. Then remove the resin by filtration using Whatman No. 1 filter paper and add the required nutrients for bacterial growth. Sterilize the medium by filtration through a sterile membrane filter (0.45 μm). Adjust the pH of the solution to pH 7.0 with sterile phosphoric acid (5 N). Pour the medium into petri dishes and allow to gel (approximately 40 min at room temperature)

Table 3. Media containing organic nitrogen for the cultivation of heterotrophic bacteria

Media	Composition (g l^{-1})		Use	Comments
Nutrient broth (pH 7.0)	Beef extract	3.0 g	Widely used as bacteriological medium	This medium can be solidified with agar (1.5%)
	Peptone	5.0 g		
Tryptone glucose yeast extract broth (pH 7.0)	Tryptone	5.0 g	Used for isolation and sporulation of *Bacillus* spp.	Media can be solidified with agar (1.5%)
	Yeast extract	2.5 g		
	Dextrose	1.0 g		
MacConkey broth (pH 7.1)	Bacto peptone	17.0 g	Selective for coliform organisms which ferment lactose in the presence of bile salts	Production of acid is detected by a color change (colorless to red). Can solidify medium with agar (1.35%)
	Proteose peptone	3.0 g		
	Sodium chloride	5.0 g		
	Lactose	10.0 g		
	Bile salts mixture	1.5 g		
	Neutral red	0.03 g		
	Crystal violet	0.001 g		
Blood agar (pH 7.3)	Beef heart infusion	500.0 g	Widely used for the cultivation of a range of pathogenic organisms (e.g. *Streptococci*)	After autoclaving, cool to 45–50°C and add 50 ml of sterile defibrinated blood. Dispense into plates
	Tryptose	10.0 g		
	Sodium chloride	5.0 g		
	Agar	15.0 g		
T-broth (pH 7.3)	Nutrient broth	8.0 g	Widely used as a growth medium in bacteriology	
	Peptone	5.0 g		
	Sodium chloride	5.0 g		
	Glucose	5.0 g		
Soil extract broth	Soil	1000.0 g	More of the heterotrophic bacteria found in soil will develop on this medium than on any other. It is nonselective in nature	Autoclave the soil in 1 l of water for 20 min at 120°C. Strain liquid and make up to volume. Adjust pH to neutrality. Medium can be solidified with agar
	Glucose	1.0 g		
	Yeast extract	5.0 g		
	Dipotassium hydrogen-phosphate	0.2 g		

Table 4. Media containing inorganic nitrogen for the cultivation of heterotrophic bacteria

Media	Composition (g l^{-1})		Use	Comments
Ayers, Rupp and Johnson's medium	Ammonium dihydrogen-phosphate	1.0 g	This medium is most frequently used as a basal media for fermentation tests	This medium can be used when a peptone-free medium is required. pH should be adjusted to 7.0 with NaOH (1 M)
	Magnesium sulfate	0.2 g		
	Glucose	1.0 g		
	Potassium chloride	0.2 g		
Eggins and Pugh's medium	Sodium nitrate	0.5 g	This medium is often used for the detection of cellulolytic bacteria	The cellulose powder should be ball-milled prior to use. The medium can be solidified with agar or silica gel
	Dipotassium hydrogen-phosphate	1.0 g		
	Magnesium sulfate	0.5 g		
	Ferrous sulfate	0.01 g		
	Cellulose	12.0 g		
de Barjac medium (pH 7.0)	Potassium nitrate	2.0 g	This medium, although not selective for denitrifiers, can be used for their growth	
	Glucose	10.0 g		
	Calcium carbonate	0.5 g		
	Potassium dihydrogen-phosphate	5.0 g		
	Sodium chloride	2.5 g		
	Ferrous sulfate	0.05 g		
	Manganese sulfate	0.05 g		
	Magnesium sulfate	0.05 g		
Noble and Graham's medium (pH 7.0)	Salicin	10.0 g	Used for the isolation of soft-rot bacteria (plant pathogens)	Soft-rot bacteria all produce acid from the salicin in peptone-free media and in the presence of bile salts
	Sodium taurocholate	5.0 g		
	Ammonium dipotassium-phosphate	1.0 g		
	Magnesium sulfate	0.2 g		
	Potassium chloride	0.2 g		
	Bromothymol blue	0.05 g		
	Agar	20.0 g		

Table 5. Media without nitrogen for the cultivation of nitrogen-fixing bacteria

Media	Composition (g l⁻¹)		Use	Comments
I	Dipotassium hydrogen-phosphate	0.8 g	Mainly used for the detection of *Clostridium*. Trace element solution contains (g l⁻¹):	Adjust medium pH to 7.2. Incubate tubes/flasks at 28–30°C under nitrogen for up to 4 weeks. Gas production suggests growth
	Magnesium sulfate	0.2 g		
	Sodium chloride	0.2 g		
	Ferrous sulfate	0.01 g	$Na_2B_4O_7$ 0.05 g	
	Manganese sulfate	0.01 g	$Co(NO_3)_2$ 0.05 g	
	Calcium chloride	0.01 g	$CdSO_4$ 0.05 g	
	Sucrose	10.0 g	$CuSO_4$ 0.05 g	
	Yeast extract	0.01 g	$ZnSO_4$ 0.05 g	
	Potassium dihydrogen-phosphate	0.2 g	$MnSO_4$ 0.05 g	
	Sodium molybdate	0.025 mg		
	Soil extract	10.0 ml	Soil extract contains: 1000 g soil in 1 l of water. Mix and autoclave at 120°C for 20 min. Strain and make up to 1 l.	
	Sodium thioglycollate	1.0 g		
	Trace elements solution	1.0 ml		
II	Dipotassium hydrogen-phosphate	0.8 g	Mostly used for the isolation of *Azotobacter*	The mannitol must be added to the autoclaved medium. A 10% mannitol solution can be filter sterilized and 20 ml added to 1 l of medium
	Sodium chloride	0.2 g		
	Ferric chloride	0.01 g		
	Magnesium sulfate	0.2 g		
	Manganese sulfate	0.01 g		
	Mannitol	2.0 g		

Cultivation of Bacteria

11

Table 6. Media used for the isolation of photosynthetic bacteria

Media	Composition (g l^{-1})		Use	Comments
I	Magnesium sulfate	1.0 g	For the isolation of sulfur-photosynthetic bacteria. Dissolve these five compounds in 700 ml water. Autoclave and cool	Trace elements solution contains (g l^{-1}): FeSO$_4$.7H$_2$O, 0.2 g; ZnSO$_4$.7H$_2$O, 0.01 g; MnCl$_2$.4H$_2$O, 0.003 g; H$_3$BO$_3$, 0.003g; CoCl$_2$.6H$_2$O, 0.02g; CaCl$_2$ 0.001g; NiCl$_2$.2H$_2$O, 0.002g; Na$_2$MoO$_4$. 2H$_2$O, 0.5 g. Autoclave, cool and add
	Calcium sulfate	0.5 g		
	Calcium carbonate	0.05 g		
	Ferrous sulfate	0.01 g		
	Ammonium sulfate	0.5 g		
	Trace elements solution	10.0 ml		
	Dipotassium hydrogen-phosphate	1.0 g (in 100 ml)	Add to above cooled medium Adjust medium to:	Grown factor solution contains (g l^{-1}): biotin, 0.001 g; Ca-pantothenate, 0.001 g; *p*-aminobenzoic acid, 0.01g; nicotinic acid, 0.02 g; thiamine hydrochloride, 0.02 g. Filter sterilize before adding to medium
	Grown factor solution	10.0 ml	pH 6.8 for *Chlorobium* pH 7.0 for *Chromatium*	
	Sodium sulfite	0.05 g	Add 0.5 g Na$_2$S for green/ purple species	
II	Ammonium chloride	0.5 g	This media is used for the isolation of nonsulfur-photo-synthetic bacteria. Dissolve the KH$_2$PO$_4$ in trace elements/ growth factor solutions, filter, sterilize and add to medium. Adjust to pH 6.8 before autoclaving	Both trace elements and growth factor solutions are as above
	Magnesium sulfate	0.4 g		
	Sodium chloride	0.4 g		
	Calcium chloride	0.05 g		Agar can be added if required
	Succinate	1.0 g		
	Yeast extract	1.0 g		
	Potassium dihydrogen-phosphate	1.0 g		
	Trace elements solution	10.0 ml		
	Growth factor solution	10.0 ml		

Table 7. Media used for the isolation of chemosynthetic bacteria

Media	Composition (g l^{-1})		Use	Comments
Thiobacillus sp.[a]	Sulfur	10.00 g	For the isolation of aerobic organisms adjust the pH of the medium as follows:	Iron solution contains (g l^{-1}): $Fe_2SO_4 \cdot 7H_2O$, 1.0 g; Na-EDTA, 1.0 g
	Ammonium sulfate	3.0 g	pH 5 for *T. thio-oxidans*	
	Calcium chloride	0.25 g	pH 7.5 for *T. thioparus*	Trace salts contain (g l^{-1}):
	Dipotassium hydrogen-phosphate	3.0 g	pH 4.5 for *T. ferro-oxidans*	H_3BO_3, 0.02 g; $CaSO_4 \cdot 5H_2O$,
	Magnesium sulfate	0.5 g		0.15 g; $(NH_4)_6Mo_7O_{24} \cdot 4H_2O$, 1.0 g
	Ferrous sulfate	0.01 g		
	Calcium chloride	0.25 g		
Nitrifying bacteria	Sodium nitrite	0.5 g	For the isolation of *Nitrosomonas* (oxidizes ammonium to nitrite) and	
	Dipotassium hydrogen-phosphate	1.0 g	*Nitrobacter* sp. (oxidizes nitrite to	
	Magnesium sulfate	0.02 g	nitrate). $NaNO_2$ can be replaced by	
	Sodium chloride	0.5 g	$(NH_4)_2SO_4$, 0.1 g; $MnSO_4 \cdot 7H_2O$,	
	Iron solution	1.0 ml	0.15 g; $COCl_2$, 0.01 g; Na-EDTA, 1.0 g	
	Trace salts solution	1.0 ml		

Both of these media may be solidified with silica gel

[a] Growth of *Thiobacillus* sp. is slow and may be detected by acid production rather than by turbidity. Phenol red (0.5 mg l^{-1}) is a useful indicator. Growth of these organisms can be inhibited by volatile compounds from rubber.

Cultivation of Bacteria

13

Table 8. Methods involving the removal of oxygen from cultures

Method	Description
Stab and shake culture	Many anaerobes can grow in deep stab or shake cultures in glucose agar. The method is very useful for the cultivation of micro-aerophilic species. A seal of liquid paraffin can be used to maintain anaerobic conditions (e.g. the Hugh and Leifson fermentation test). However, tests have shown that oxygen can diffuse through the paraffin
Anaerobic jar culture	If surface cultures are required, the plates should be placed in an anaerobic jar. Older jars have inlet and outlet valves which allow the air in the jar to be replaced with hydrogen after evacuation of the oxygen. When complete the valves are closed. A palladium catalyst allows the combination of hydrogen with oxygen resulting in a partial vacuum. Newer jars are made of plastic and require the use of Gas-Pak® which is a sachet to which water is added and it is then added to the jar. The chemical reaction uses oxygen and generates both hydrogen and carbon dioxide (this may stimulate growth of some *Clostridia*). An additional strip of paper containing methylene blue indicator is added to the jar. The paper turns blue in the presence of oxygen but remains white in the absence of oxygen
Pyrogallol cultures	Cultures can be kept in a desiccator with a sufficient volume of alkaline pyrogallol to remove all the oxygen from the atmosphere above the cultures. After placing the pyrogallol and cultures in the desiccator, KOH should be poured in quickly and the lid placed in position. Alkaline pyrogallol absorbs CO_2. This may inhibit some *Clostridia* but can be overcome using carbonates, although O_2 absorption is much slower. Hydrogen inhibits nitrogen fixation and is therefore unsuitable for culturing nitrogen-fixing clostridia
Activated steel wool	An alternative way to absorb oxygen is with activated steel wool. Steel wool (10 g, grade 1) is dipped in an activating solution containing (l^{-1}): $Ca_2SO_4.5H_2O$, 2.5 g; Tween 80, 2.5 g, acidified with NH_2SO_4. After a few seconds drain the steel wool and place in a petri dish in an anaerobic jar. The steel wool and copper form many incomplete iron–copper couples which help absorb oxygen. CO_2 is generated by mixing equal weights of $MgCO_3$ and $NaHCO_3$. Five grams of the mixture are placed with 10 ml of water in a beaker in the anaerobic jar. CO_2 is released slowly during incubation

Table 9. Methods involving the reduction of culture medium

Method	Description
The iron nail method	The medium is heated in boiling water for 10 min, cooled and a sterile iron nail is added. The medium is depleted of oxygen due to corrosion of the nail. The medium is inoculated in the usual way
Robertson's cooked meat broth	Lean meat (1 kg) is minced and simmered in 1 l water containing 1.5 ml of NaOH (1M) for 20 min. After simmering, strain and wash thoroughly with distilled water and partially dry. Distribute into tubes to a depth of at least 2 cm. Nutrient broth is then added to a depth of approximately 10 cm and the medium is autoclaved. The sterilized meat contains reducing substances which are effective in maintaining anaerobic conditions at the bottom of the tube. The reducing activity is shown by the pink hematin at the bottom of the tube
Thioglycollate medium	Add thioglycollic acid (0.1%) to nutrient broth before adjusting the pH. Glucose (1%) must be added. The medium may be solidified with agar, but it is more usual to use a semi-solid medium (0.5% agar). The increased viscosity of the medium prevents the distribution of oxygen by convection currents. Methylene blue is added to act as an indicator. A blue–green layer on the surface shows the depth to which O_2 has diffused. Inocula should be introduced carefully by means of a fine pipette at the bottom of the tube
Reinforced *Clostridium* medium	The medium is mainly used for the cultivation and enumeration of *Clostridia* in food, pathological specimens, soil etc. The medium contains (l^{-1}): yeast extract, 3.0 g; cysteine hydrochloride, 0.5 g; sodium chloride, 0.5 g; sodium acetate, 3.0 g; agar, 0.5 g. Autoclave for 20 min
Pankhurst's medium	Dissimilatory sulfate-reducing bacteria (e.g. *Desulfovibrio*) obtain energy by reduction of sulfate which acts as the terminal electron acceptor in anaerobic respiration. Dissolve K_2HPO_4, 0.5 g; NH_4Cl, 1.0 g, $CaSO_4$, 1.0 g; $MgSO_4.7H_2O$, 2.0 g; 70% (w/v) sodium lactate solution, 50 g; oxoid ionagar no. 2, 10.0 g into demineralized water (930 ml). Adjust pH to 8.1. Autoclave for 20 min at 115°C and cool rapidly to prevent resolution of oxygen. Add supernatant (50 ml) from an aqueous solution (1% w/v) of $FeSO_4.(NH_4)_2.SO_4.6H_2O$ (steam sterilized for 1 h on 3 successive days); 10 ml of 10% (w/v) yeast extract (filter sterilized) and 10 ml sodium thioglycollate (10% w/v) filter sterilized. Make oxygen free or fill the vessel. NaCl (3%) should be added for marine strains

Cultivation of Bacteria

Chapter 2 METHODS OF BACTERIAL ISOLATION

1 Introduction

Microbiological work depends on the maintenance of pure cultures of bacteria. There are many precautions that must be observed to exclude other bacteria. Aseptic technique is generally a matter of common sense. However, the following points should always be considered when preparing and carrying out bacterial work.

1. Any surface exposed to air will quickly become contaminated. Instruments which can be sterilized by heating in a bunsen flame (e.g. inoculating loops) can be left exposed, but must be flamed before and after use. Equipment which cannot be sterilized in this way such as glassware must be wrapped, sterilized and left unopened until required. Nothing should be in contact

3 Purification of bacteria from contaminated cultures

The basis of this isolation technique is the progressive dilution of a suspension containing more than one bacterial species. Before deciding that a culture is pure, using either method, colonies should be picked off, grown and re-separated until all the colonies are the same. Use staining as a check for purity on the final isolations. A description of the methods used to purify bacteria from contaminated bacteria is shown in *Table 2*.

4 Isolation of bacteria from natural sources

The aim of many viable counting methods is to estimate the number of living bacteria in a sample. To do this a medium with nonsterile surfaces during use. Plugs and tops of flasks and tubes must not be placed on the bench, and sterile containers must not be left open.

2. Bacteria in the air are difficult to destroy. Germicidal sprays and UV light reduce the number of bacteria, but even when using these methods good aseptic technique must always be maintained. The same is true for laminar flow cabinets.

3. Provision must always be made for the disposal of used equipment so that it does not contaminate other material. Buckets containing disinfectant should be available and disposable plastics should be autoclaved before disposal.

2 Sterilization

Media are usually sterilized after distribution into bottles, tubes or flasks. Sterile media may be transferred aseptically into previously sterilized containers. *Table 1* lists the methods commonly used for the sterilization of media.

satisfying the nutritional requirements of as many bacteria in the sample as possible is required. Other types of viable count and isolation techniques employ the reverse principle; (i.e. to encourage only one type of bacteria to grow at the expense of all the others present in a sample). In samples from natural sources (e.g. soil) the bacteria to be isolated may only be present in small numbers. The first step therefore is to obtain an enrichment culture by one of the following methods:

1. Using selective media.
2. Using selective conditions of incubation.
3. Selective pre-treatment of bacteria.

Several generations of subcultures on liquid or solid media may be necessary, but the final step will consist of plating out the bacteria using one of the methods described earlier.

5 Use of selective media

A completely selective medium allowing the growth of only a

Bacterial Cell Culture

single species cannot be found, but media can be produced which will discourage growth of all but the required bacteria. Some selective media function in a different fashion. With this type of selective media, certain bacteria exhibit distinct biochemical or morphological characteristics which allow them to be recognized easily. This may be very important when the range of species is small (e.g. medical microbiology). When a large number of species is involved (e.g. soil) differential media cannot be used to separate taxonomic groups, but may be useful to distinguish between bacteria with different biochemical properties such as chitin degradation. Selectivity may be achieved in one of three ways:

1. Addition of compound to the medium which discourages growth of all but the required species.
2. Altering the pH of the medium.
3. Omission of a compound which is required by most bacteria, but not the ones selected for.

Table 3 lists the methods used for the production of selective media.

6 Use of selective incubation

Selective incubation of cultures can select quite different portions of a bacterial population. *Table 4* describes the use of temperature and aeration in the selective isolation of bacteria.

7 Selective pre-treatment of material

The sample from which isolations are made can be pre-treated by physical or chemical means. Using any of these methods, the enriched material is plated out on a selective medium for the particular organism. A good enrichment technique should result in 90% of the organisms present being the ones required.

Table 1. Methods of sterilization

Treatment	Comments
Autoclave	Most media can be sterilized by treatment with steam under pressure in an autoclave (usually a 15–20 min treatment at a pressure of 1 kg cm^{-2}). This raises the steam temperature to 121°C
Steaming	Media which cannot be autoclaved satisfactorily (e.g. sugar media) may be sterilized by intermittent steaming. Samples are heated over boiling water in a steamer (85–95°C) for 15 min on each of 3 successive days. Between treatments the material must be left at a suitable temperature for the growth of endospores
Filter sterilization	It is necessary to sterilize some ingredients of a medium separately and then add them to the medium before use. Heat-labile ingredients (some antibiotics, sugars and vitamins) are sterilized by filtration through a membrane filter (pore size 0.2–0.45 μm) which excludes bacteria
Dry heat	Dry glassware such as empty flasks, pipettes etc. cannot be satisfactorily autoclaved and must be sterilized in a hot-air oven for a minimum of 2 h at a temperature of 160°C. All material should be allowed to cool to room temperature before removal from the oven
Other methods	Sterilization of natural materials (e.g. soils, seeds etc.) is difficult as heat sterilization can alter the material. Radiation can be used and causes minimal changes. Plant material can be surface sterilized with microbiocides such as HgCl$_2$, but chemicals must be carefully removed before use

Bacterial Isolation

19

Table 2. Methods of isolation of bacteria from contaminated cultures

Method	Description
Pour plate	Inoculate one tube with one loopful of the suspension and mix by rotating it between the hands. Transfer one loopful of the mixture from the first tube to the second and mix. Repeat for the third tube. Pour the contents of all the tubes into sterile petri dishes and allow to set. The second and third dilutions should show well separated colonies after incubation. If the suspension is very heavy, initial dilution in sterile saline may be required
Streak plate	This is a quicker, though less reliable method. Prepare a plate of solid medium. Dry the plates for 30 min before use at 45°C. Place a small drop of suspension on the agar near the edge of the dish. Draw this out into a single broad streak. After sterilizing the loop draw a set of streaks cutting across the first streak and covering one half of the plate. Turn the plate through 90° and continue the process. After incubation well separated colonies should be found along the streak marks

Table 3. Methods of producing selective media for the isolation of bacteria

Method of selection	Example	Description
Addition of substrates	MacConkey's bile-salt-lactose broth. Use to detect coliform organisms in milk and water	Bile salts discourage the growth of most bacteria except those of intestinal origin. The presence of 1% lactose and a pH indicator makes the medium differential. Coliform organisms will produce acid and gas from the medium. When solidified the

	same medium can be used to detect the typhoid–dysentery group of bacteria. Coliforms produce acid, *Salmonella* and *Shigella* do not
Potassium selenite (0.04%)	Inhibits growth of coliforms
Sodium azide (0.25%)	Used to inhibit Gram-negative bacteria, allowing isolation of fecal streptococci
Potassium tellurite (0.002%)	Inhibits growth of coliforms and pseudomonas; used in blood agar to isolate *Corynebacterium diptheriae*
Crystal violet (0.001%)	Inhibits Gram-positive bacteria; used for isolating *Pseudomonas*
Actidione (0.005%)	Inhibits yeast and fungi; used to enumerate bacteria in mixed floras
Penicillin (1000 units ml^{-1})	A variety of antibiotics are available to inhibit a number of group of organisms
Streptomycin (50 µg ml^{-1})	
Alteration of pH	
Dieudonne's blood–alkali agar	Organisms capable of growth at high pH (e.g. *Vibrio* and *Pseudomonas*) can be selected for by adjusting the pH to 9–10. Acid-tolerant bacteria can be isolated if the medium pH is low (pH 5). Lactobacilli are isolated on media containing 0.5% acetic acid. After isolation, the bacteria should be grown on normal media
Omission of substances	
Mannitol phosphate agar	Media with inorganic substances are used to isolate autotrophic bacteria. Several transfers are required as some organic matter is introduced with the inoculum. Mannitol phosphate media, used for nitrogen fixing organisms, is an example of this method of selection

Bacterial Isolation

Table 4. Methods of producing selective incubation for the isolation of bacteria

Method of selection	Example	Description
Temperature	Isolation of thermophiles	Bacteria have definite temperature requirements with maximum temperatures above which they fail to grow, and minimum temperatures below which they are unable to grow. Thermophiles grow at 50°C. Mesophiles rarely grow above 45°C and psychrophiles seldom grow above 25°C. These limits can be altered by various cultural conditions but the temperature range of a bacteria is easy to establish. To isolate such organisms from soil, a moist sample should be incubated at 50°C or higher for 3 days. A suspension of this should be made in 25% Ringer's solution and heated in a water bath for 20 min at 100°C. Plate out 1 ml of this suspension on an agar plate and incubate at 50°C, with the plates in a polythene bag to prevent drying
Aeration	Isolation of obligate anaerobes	Bacteria have a definite relationship to aeration conditions. Some require oxygen (obligate aerobes) while others only grow in the absence of oxygen (obligate anaerobes). By changing the aeration conditions of incubation, different groups of bacteria can be selected

Table 5. Methods of producing selective pre-treatment of material for the isolation of bacteria

Method of selection	Example	Description
Physical methods	Isolation of spore-forming bacteria	Bacterial spores are heat resistant (thermoduric) to varying degrees. A few species can tolerate exposure at 100°C for 20 min but most will survive treatment at 70–80°C for 10 min. More species will survive in a medium containing colloidal organic matter than in pure water. To obtain spore-forming bacteria from a mixed suspension containing nonsporing species, incubate the suspension in a water bath at 75°C for 10 min prior to plating out. *Bacillus* species will grow on plates incubated aerobically, *Clostridium* species grow on plates incubated anaerobically
Chemical methods	Isolation of cellulolytic bacteria	Static enrichment cultures may be produced by mixing a quantity of the specific substrate being studied (e.g. cellulose) with soil and incubating for several days. The length of incubation will be dependent on the complexity of the substrate and the original inoculum of organisms present. A suspension of this material provides the starting point for isolation
Perfusion enrichment cultures	Isolation of decomposing bacteria from soil	This method is based on a column containing soil mixed with the substance being investigated. Sterile, aerated liquid medium (with or without the substance) is percolated through the column and then recycled. The decomposition of the compound can be followed using this method, and the organisms responsible for decomposition isolated. One disadvantage of this system is the artificial nature of the incubation conditions

Continued

Bacterial Isolation

Table 5. Methods of producing selective pre-treatment of material for the isolation of bacteria, *continued*

Method of selection	Example	Description
Continuous culture enrichment	Used to isolate a range of organisms	Continuous culture in a chemostat can also be used to isolate bacteria. In this system, the rate of bacterial growth is dependent on the rate of nutrient addition. All but one of the required nutrients are present in excess and so only one nutrient is limiting the amount of growth possible. The rate of flow of the medium, and therefore the growth rate, is determined by a medium pump which can be set at a variety of values. Using this system, bacteria can be grown for long periods of time at very low substrate concentrations using a variety of limiting factors. Therefore, different bacteria may be isolated using this system, than could be isolated in batch cultures, which are characterized by initial high substrate concentration which then falls quickly.

Chapter 3 MICROSCOPIC EXAMINATION OF BACTERIA

1 Examination of unstained bacteria

Some of the major characteristics of bacteria are morphology size, shape, arrangement and structure. Since the average size of a bacteria is approximately 2–4 μm in length and 0.5–1.0 μm in diameter, any bacterial examination must take place using high magnification ($\times 1000$). To reach these values of magnification, use is made of oil immersion lenses with focal lengths of around 2 mm. Always locate bacteria on a slide using the lower magnification lenses (e.g. $\times 10$, $\times 40$) before examination using oil immersion lenses ($\times 100$). When bright-field microscopy is used for the examination, it is preferable that the cells be stained to make them more readily visible. Unstained bacterial cells are practically transparent and are best observed by techniques that permit special and critical control of illumination such as phase contrast and dark-field illumination. The most useful and frequently used method of examining live bacteria is the hanging drop method (*Table 1*).

2 Methods of staining bacteria

The preparation of films for staining is as follows:

1. Good bacterial films can be made by removing a small quantity of surface growth from a solid medium and mixing it with water. The resultant suspension should be faintly opalescent.

2. A drop of this liquid should be placed on a slide with a sterile loop. The drop should be spread evenly and allowed to air dry. This should take no more than 1 min. When dry, the film should just be visible.

Microscopic Examination of Bacteria

3. Because of their small size, bacteria dry without great distortion and so the only fixation required is to pass the slide quickly two or three times through the bunsen flame.

2.1 General staining principles

1. Dyes may be either acidic, basic or neutral in type. Basic dyes have the greatest affinity for cell nuclei because of the acidic nature of nuclear material. Acid dyes have greatest affinity for the cytoplasm. The most commonly used dyes are salts. A salt is composed of a positively charged ion (cation) and a negatively charged ion (anion). Dyes are salts containing both an organic and an inorganic ion.

2. When staining always use positive and negative control bacteria.

3. Stains should always be poured directly on to the slide. Always flood the entire slide with the stain. When staining is complete, the stain should be removed with running water and the slide blot dried with filter paper.

2.4 Gram's method

Some bacteria, when treated with *para*-rosaniline dyes and iodine retain the stain when treated with a decolorizing agent such as alcohol or acetone, while other bacteria lose the stain. Gram-negative forms, which are those that lose the stain, can be counterstained. The difference between positive and negative forms is caused by differences in the structure of cell walls. Gram-positive forms probably have cell walls containing mainly peptidoglycan which prevent the stain being leached out from the cytoplasm with alcohol. This stain is the most important differential stain used in bacteriology for characterizing bacteria. A useful form of the Gram stain is Hucker's modification (*Table 3*).

2.5 Acid-fast stain (modified Ziehl–Neilsen method)

Some bacteria (e.g. *Mycobacterium*) cannot be stained by ordinary aniline dyes except by using phenol and sometimes heat. These bacteria have waxy cell walls because they contain large quantities of lipoidal material. Such walls are hydrophobic and are impermeable to stains and other chemicals in

2.2 Simple stains

Staining of a bacterial film, called a smear, may be performed to reveal size, shape and arrangement of the cells. The cells are stained by the application of a single staining solution. The process is called a simple stain and examples of simple stains commonly used for demonstration of general morphology are given in *Table 2*.

2.3 Differential staining

It is possible to acquire additional information about the morphology and chemical composition of bacteria through the use of differential staining. There are more complex staining procedures designed to demonstrate different parts of the bacterial cell, and different types of bacteria. Differential staining procedures usually require the treatment of bacterial smears with a number of reagents. The appearance of cells following treatment may distinguish between two different bacterial types on the basis of the color they retain. Careful examination of appropriately stained bacteria provides invaluable information about the morphology of the bacteria.

aqueous solutions. In order to stain these bacteria special staining procedures must be employed. When stained, these bacteria resist decolorization with acids and are called acid-fast bacteria (*Table 4*). This stain is used to detect *M. tuberculosis*.

2.6 Spore stain (Conklin's method)

In response to unfavorable environmental conditions, some bacteria such as *Bacillus* and *Clostridium* develop a spore which is resistant to physical and chemical agents. Spores are not easily stained by Gram's stain and are therefore frequently observed as clear areas in an otherwise stained cell. This is because spores are more acid-fast than the rest of the cell. However, once stained the spore does not easily release the stain. The easiest method of staining spores is the Conklin's method (*Table 5*).

2.7 Flagella stain

Motile bacteria possess one or more very fine thread-like appendages called flagella. Flagella are coiled into rigid spirals that revolve around their points of attachment. Although the length of flagella may be many times the length of the bacteria

Microscopic Examination of Bacteria

itself, the diameters of flagella are only 0.01–0.05 μm. Flagella are therefore below the size limit for light microscopy. However it is possible to mordant the flagella before staining so their apparent size is increased. Also, flagella are easily lost. Care must be taken at all stages. Bacteria for flagella staining should be grown on moist agar slopes. Check for motility using the hanging drop method. The common methods used to stain flagella are given in *Table 6*.

2.8 Capsule staining

The cell wall of many species of bacteria is surrounded by a polymeric substance referred to as a capsule or slime layer. The size of the capsule varies with the species and the composition of the growth media. Chemically the capsular material may be a polysaccharide, a glycoprotein or a polypeptide. Among pathogenic bacteria, the presence of a capsule is an indication of a virulent form of the bacteria. This is because the capsule protects the cell from being phagocytized by the host white blood cells. Capsules may be easily confused with other structures or with artifacts.

Anthony's method has the advantage that it stains the cell and capsule different colors. A range of stains used to stain the bacterial capsule is given in *Table 7*.

2.9 Staining of other bacterial structures

It is possible to specifically stain a number of other bacterial structures using a range of stains and methods. Staining protocols using a range of stains and methods. Staining protocols are presented in *Table 8*. Among the structures that can be stained are cell walls, nuclear material and rhizobia.

2.10 Fluorescent staining

Fluorescent stains are useful for the examination of bacteria in water and soil samples, as fluorescing cells often contrast well with background material. They are also useful for identifying particular biochemicals in cells (see following section). Short-wavelength blue light can be used to make these compounds fluoresce, but ethidium and propidium bromides fluoresce well under green light. Acridine orange fluoresces well under UV light. Examples of fluorescent stains are given in *Table 9*.

Table 1. Examination of living, unstained bacteria using the hanging drop method

Step	Method
1	Place a small piece of rolled Plasticine® on a slide so that it forms a circle with a similar diameter to a coverslip
2	Using a wire loop, place a drop of bacterial suspension on a coverslip
3	Invert the slide and gently press the Plasticine® on the coverslip. Quickly invert the slide, leaving the bacterial suspension as a 'hanging drop'
4	Examine the slide under a × 10 objective, focusing on the edge of the drop. Once focused, view with the × 40 objective

This method is used to detect motility, but this must be distinguished from Brownian movement, a phenomenon exhibited by any small particle in an aqueous suspension. Examination under oil immersion is difficult since currents of liquid are set up during focusing. Phase contrast microscopy is the easiest method for the examination of unstained bacteria. Bacteria appear as dark objectives against a pale background. This method, using special objectives and condensers, depends upon the change of phase of white light as it passes through objects with different refractive indices.

Table 2. Composition and description of a range of simple stains

Stain	Composition		Description
Loeffler's alkaline methylene blue	Methylene blue chloride	1.6 g	Dissolve the methylene blue in ethanol and then add the KOH
	Ethanol, 95% (v/v)	100.0 ml	Flood the bacterial film for 10 sec, wash with water and blot dry.
	KOH, 0.01% (w/v)	100.0 ml	This stain can be used to detect metachromic granules
Ziehl's carbol fuschin	Saturated alcohol-soluble basic fuschin	10.0 ml	Flood the bacterial film with stain for 20 sec, wash and blot dry.
	Phenol (5%)	100.0 ml	An intensively stained preparation is obtained without differentiation

Microscopic Examination of Bacteria

Table 3. Composition and description of the Gram stain

Stain	Composition		Description
Gram's stain (Hucker's modification)	Ammonium oxalate crystal violet		Stain the film with ammonium oxalate crystal violet solution for 1 min. Wash with tap water for 2–3 sec
	Ammonium oxalate	0.8 g	
	Crystal violet	2.0 g	
	Ethyl alcohol	20.0 ml	
	Distilled water	80.0 ml	
	Lugol's iodine solution		Immerse slide for 1 min in Lugol's iodine solution. Wash slides and blot dry. Treat with 95% ethyl alcohol for 30 sec
	Iodine	1.0 g	
	Potassium iodide	2.0 g	
	Distilled water	300.0 ml	
	Saffranin (10–30 sec)		After washing with water, counter stain with saffranin solution for 10–30 sec
	Saffranin (2.5% solution in 95% ethanol)	10.0 ml	
	Distilled water	100.0 ml	

The preparation is washed, blotted and examined under an oil-immersion lens. Gram-positive bacteria will be stained purple, Gram-negative bacteria will stain red. Organisms with known reactions should be used as controls on the same slide. Always look at several fields of view when trying to interpret the results

Table 4. Composition and description of the acid-fast stain

Stain	Composition		Description
Acid-fast staining (modified Ziehl–Nielson's method)	Ziehl's carbol fuchin[a] (5 min with heat)		Stain the film with Ziehl's carbol fuchsin for 5 min applying heat to give steam. Do not let slide dry out. Wash the slide with water
	Basic fuchsin	0.3 g	
	Ethyl alcohol	10.0 ml	
	Phenol (heat melted crystals)	5.0 ml	
	Distilled water	95.0 ml	
	Decolorizing solution		After removing excess water, treat the slide with decolorizing solution. Wash the film immediately with water and repeat decolorizing step until the film appears faintly pink
	Hydrochloric acid	3.0 ml	
	Ethyl alcohol (95%)	97.0 ml	
	Counter stain		Counter stain with aqueous methylene blue for 20–30 sec. Wash and blot dry
	Methylene blue	0.3 g	
	Distilled water	100.0 ml	

Acid-fast organisms retain the red stain while other bacteria are stained blue. Endospores of *Bacillus* are also acid-fast (see below)

[a]Dissolve the fuchsin in the ethyl alcohol and then add the phenol dissolved in water. Mix and stand for a few days. Filter.

Microscopic Examination of Bacteria

Table 5. Composition and description of the spore stain

Stain	Composition		Description
Spore staining (Conklin's method)	Aqueous malachite green (5 min with heat)		Stain slide with aqueous malachite green for 5 min, heating the slide gently until it steams. Wash slide with water for 30 sec
	Malachite green	5.0 g	
	Distilled water	100.0 ml	
	Counter stain (aqueous saffranin)		Counter stain with saffranin for 30 sec
	Saffranin (2.5% solution in 95% alcohol)	10.0 ml	
	Distilled water	100.0 ml	

Wash and blot dry slide. Under oil immersion, the spores are stained green and the rest of the cell red

Table 6. Composition and description of a range of flagella stains

Stain	Composition	Description

A small drop of the active bacteria should be transferred to a clean slide and allowed to run the length of the slide

Modified Rhode's stain	Safford and Fleisher's mordant (25% saturated aqueous)		Allow the film to dry naturally and fix with gentle heat. Flood the film with mordant for 3–5 min, using a fresh, filtered solution. Wash slide with demineralized water
	Picric acid	10.0 ml	
	Tannic acid	5.0 g	
	Ferrous sulfate	7.5 g	
	Distilled water	90.0 ml	
	Sensitized silver nitrate solution[a]		Drain and treat with hot sensitized silver nitrate solution
	Silver nitrate	2.0 %	
	Ammonia	0.88 M	

The mordant swells the flagella and the mordanted flagella cause the silver nitrate to be reduced. Thus the flagella become silver plated. Wash the slide and blot dry carefully. Examine under oil immersion. Flagella appear as pale gray or brown threads against a colorless or pale brown background

Leifson's stain	Tannic acid solution		Prepare 3 stock solutions: tannic acid, salt and fuschin. Allow several hours to dissolve and filter before use. Pipette 1 ml of the mixed solutions then filter. Mix equal portions of the three solutions on to the slide and stain for 6–10 min. Rinse under a slow running tap. Drain and dry in air. Flagella will stain red
	Tannic acid	3.0 g	
	Phenol	0.2 g	
	Distilled water	100.0 ml	
	Salt solution		
	Sodium chloride	1.5 g	
	Distilled water	100.0 ml	
	Fuchsin		
	Basic fuchsin	1.2 g	
	95% alcohol	100.0 ml	

[a]To the silver nitrate solution add ammonia until the precipitate dissolves. Add more silver nitrate until the solution becomes faintly turbid.

33   *Microscopic Examination of Bacteria*

Bacterial Cell Culture

Table 7. Composition and description of a range of capsule stains

Stain	Composition		Description
Anthony's method	Acetic crystal violet		Films are stained in crystal violet for 5 min. Wash the slides with copper sulfate solution, blot dry and examine
	Crystal violet	0.1 g	
	Glacial acetic acid	0.25 ml	
	Distilled water	100.0 ml	
	Copper sulfate solution		
	$CuSO_4.5H_2O$	20 g	
	Distilled water	100.0 ml	

The capsules are stained blue–violet, the cells dark blue

Negative staining	Water soluble nigrosin	10.0 g	Dissolve nigrosin in distilled water and boil for 30 min. Add the formalin. Filter the resulting solution twice through double filter paper. To carry out the stain, a loopful of culture is mixed with an equal volume of stain on a slide. The mixture may be examined wet or dry
	Distilled water	100.0 ml	
	Formalin	10.5 ml	

The background should be gray and the capsules seen as a clear area

Table 8. Composition and description of a range of stains

Stain	Composition		Description
Nuclear material (Robinow's stain)	Ethanol solution		Make a suspension of an 18 h culture of bacteria in a few drops of sterile water. Inoculate the surface of a freshly poured plate of nutrient agar which has been dried at 37°C for 30 min. Incubate the plate for 2–5 h. Remove a block of agar, place it on a coverslip and place in a dish containing osmium tetroxide vapor. Keep the plate tightly closed. After 5 min fixation, place the agar block, face down, on a second coverslip and slide the agar off. Let the impression film dry, and store in 60% alcohol until required. Remove the preparation and wash in water. Place in 1 M HCl for 10 min at 60°C. Wash the coverslip three times in water and stain with 2–3 drops of Gurr's stain for 30 min. Invert the coverslip on a small drop of water on a slide and examine under oil immersion
	Ethanol (100%)	60.0 ml	
	Distilled water	40.0 ml	
	Gurr's Giesma stain		
	Gurr's stain	1 drop	
	Sorenson's buffer (pH 7.0)	1.0 ml	

The deeper colours (blue and violet) are localized in the chromatinic material

Cell walls stain	Tannic acid solution		A film is prepared in the same way as described for the nuclear material stain. Treat the film with tannic acid solution for 30 min. Wash with water and then stain with crystal violet solution
	Tannic acid	5.0 g	
	Distilled water	100.0 ml	
	Crystal violet solution		
	Crystal violet	0.2 g	
	Distilled water	100.0 ml	

The cell wall will stain purple

Continued

Microscopic Examination of Bacteria

Table 8. Composition and description of a range of stains, *continued*

Stain	Composition		Description
Barlow's stain for rhizobia	Mercuric chloride solution		Crush a root nodule (which has been sterilized with $HgCl_2$ for 30 min and washed in a little water). Transfer a loop of the turbid solution to a slide and allow to dry Flood the slide with water, and while wet add a drop of Barlow's stain at one end of the slide. Allow the stain to diffuse through the smear and then wash the slide in running water and allow to dry. Examine under oil immersion
	$HgCl_2$	0.1 g	
	Distilled water	100.0 ml	
	Barlow's stain (boil and cool)		
	Glucose	50.0 g	
	Glycerol	50.0 g	
	Gentian violet	3.0 g	
	Distilled water	50.0 ml	

Table 9. Fluorescent stains used in the microscopic examination of bacteria

Stain	Composition		Description
Fluorescein isothiocyanate (stains protein)	0.5 M carbonate–bicarbonate buffer (pH 7.2)	1.3 ml	Stain slides for 1 min at 37°C, rinse for 10 min in 0.5 M carbonate–bicarbonate buffer (pH 9.6) and mount in buffered glycerol (pH 9.6)
	0.85% physiological saline	5.7 ml	
	Crystalline fluorescein isothiocyanate (mix and leave for 10 min)	5.3 mg	
	0.5 M carbonate–bicarbonate buffer (pH 9.6)	10.0 ml	
	Glycerol (pH 9.6)	10.0 ml	
Observations are best made under short-wavelength blue light			
Acridine orange	Acridine orange	1.0 mg	Stain the slide with acridine orange for 2 min and remove by rinsing in tap water
	Distilled water	15.0 ml	
Specimens are best irradiated with UV light. Bacteria may stain either green (RNA) or orange (DNA), although dead cells and Gram-positive cells may stain orange			
Ethidium bromide	Ethidium bromide	5.0 mg	Soil smears can be stained with ethidium bromide and then washed for 1 min, and mounted in distilled water. Alternatively cells from water samples filtered through nitrocellulose acetate membranes can be stained, as can agar films
	Distilled water	10.0 ml	

The fluorochrome can be excited with 390–490 nm light, when it fluoresces red, although some soil particles may autofluoresce. Less autofluorescence is observed if the specimen is irradiated with 520–560 nm green light, and this has the added advantage of allowing double staining with fluorescein isothiocyanate to be carried out

Microscopic Examination of Bacteria

Chapter 4   ESTIMATING POPULATION SIZES OF BACTERIA

1 Bacterial biomass measurement

Biomass is usually calculated in terms of the dry weight of bacteria per ml of a suspension and may be determined directly using the total mass method described in *Table 1* provided that the suspension does not contain additional material. Alternatively, protein can be used as an estimation of bacterial biomass. One method commonly used to estimate protein concentration is the Lowry method which is also described in *Table 1*.

2 Turbidometric estimations

When large numbers of determinations are to be made (i.e. when making growth curves) turbidometric estimations can be made. Turbidity is the effect of light scattering by a colloidal

A plot of log (I_o/I), the density or extinction of the suspension, against c, the bacterial concentration, gives a straight line with a slope of kl. The equation only holds true for a limited range of concentrations. When scattered light is measured, its intensity depends on the angle between the axis of the incident light and that of the receiver, due to the differing degrees of scattering at different angles. Bacteriologists follow changes in turbidity using either a nephelometer or a spectrophotometer. The nephelometer permits the measurement of scattered light from suspensions of extremely low turbidity, and is preferable to measuring undeviated light by means of a spectrophotometer, which is less sensitive at low cell concentrations, although it is more sensitive at higher concentrations.

3 Theory of bacterial growth in batch culture

The growth of a bacterial culture undergoing binary fission is expressed by the equation:

$$x = x_o . 2^n$$

where x = the final number of bacteria; x_o = the initial number of bacteria; n = the number of generations.

If n generations are produced in time t then:

$$g = \frac{t}{n} \text{ and } n = \frac{t}{g}$$

where g = the generation or doubling time.

Therefore

$$x = x_o . 2^{t/g}$$

and

$$\ln x = \ln x_o + \ln 2 . t/g$$

Estimating Population Sizes of Bacteria

suspension. Many microbial suspensions are colloidal because the cells do not settle out of suspension quickly. These suspended cells scatter light, producing turbidity. As microbial cells grow, the number of cells increases, accompanied by a proportional increase in turbidity. These measurements are meaningless, however, unless translated into terms of cell concentration (viable or total counts). However, the relationship between turbidity and cell concentration is different for each species. Turbidity readings obtained for a series of dilutions of a known number of organisms should be plotted as a calibration curve.

Incident and scattered light are related by an equation resembling that derived from the Lambert–Beer law:

$$\log (I_o/I) = klc$$

where I_o = the incident light intensity; I = the transmitted (not scattered) light intensity; k the extinction coefficient; l = the depth of suspension; c = the bacterial concentration.

this is usually written

$$\ln x = \ln x_o + \mu t$$

where

$$\mu = \ln 2/g = 0.6931/g$$

μ represents the number of e-fold increases in cell numbers that have occurred in time t.

If the equation $\ln x = \ln x_o + \mu t$ is differentiated, it becomes the familiar growth equation:

$$\frac{dx}{dt} = \mu x$$

which may be stated as the rate of growth of the population being proportional to the initial number of organisms present.

4 Counting methods

Counts may be used to determine the number of organisms in a sample and these counts can be converted to estimates of

capable of growth and multiplication in the material. Aseptic methods must be used throughout. In all these methods the preparation of dilutions is the first step. A weighed amount or measured volume of the material being examined is suspended in a known volume of sterile diluent, mixed well by shaking, and further dilutions made from this. Sometimes, the weight of the cells is estimated by evaporating the suspension to dryness at 105°C and weighing the residue. *Table 3* describes the various techniques used to estimate viable bacterial counts.

The diluent is most important. Sometimes the natural milieu is used (e.g. seawater) but more often sterile 25% Ringer's solution is recommended. This contains:

NaCl	9.0 g
KCl	0.42 g
$CaCl_2$	0.48 g
$NaHCO_3$	0.2 g
Distilled water	4.0 l

biomass if suitable conversion factors are known.

4.1 Total counts

Total counts include the various methods of estimating the total number of bacteria per unit volume or weight of a sample, without distinguishing between viable and nonviable organisms. *Table 2* describes a number of techniques commonly used in bacteriology to estimate total counts.

4.2 Viable counts

These methods aim to estimate the number of bacteria

Tenfold dilutions from viable counts are made by placing measured quantities of sterile diluent in sterile tubes using aseptic techniques. Measured volumes cannot be sterilized by heat because of consequent volume changes.

Repeat until the sample is sufficiently diluted. The number of dilutions needed will depend on the material concerned with only the two or three highest may be used. For tap water 10^{-1} and 10^{-2} are required; for pasteurized milk 10^{-2} and 10^{-4} are used; for contaminated water up to 10^{-4} may be required and for soil between 10^{-6} and 10^{-9}.

Table 1. Estimation of bacterial biomass

Method	Chemicals		Description
Total mass	Formalin	1%	A sample of culture is treated with formalin to give a final concentration of 1% (v/v) and centrifuged in a weighed tube. The pellet is washed with 0.85% (w/v) saline containing 1% formalin and then in 0.05% saline and finally distilled water. The pellet is dried to a constant dry weight at 105°C, placed in a dessicator over phosphorus pentoxide and allowed to cool and then weighed in a balance
	Saline	0.85%	
	Saline	0.05%	
	Distilled water	10 ml	
Protein content (Lowry's method)	Solution A		Place 3 ml of solution C and up to 0.6 ml of sample (containing 15–300 mg protein) in a 10 ml tube. Mix, then stand at room temperature for 10 min to dissolve the sample. Add 0.3 ml Folin's reagent and mix at once. After exactly 30 min measure the absorbance in a spectrophotometer at 750 nm. Crystalline bovine albumin can be used as a standard. If the sample does not dissolve in solution C, treat it with 0.1 ml of 1 M NaOH and after 30 min add 1 ml of Folin–Ciocalteau reagent
	Na_2CO_3	2.0 g	
	0.1 M NaOH	100.0 ml	
	Solution B		
	$C_4H_4KNaO_6.4H_2O$	1.0 g	
	Distilled water	100.0 g	
	Solution C		
	Solution A	50.0 ml	
	Solution B	1.0 ml	
	Folin–Ciocalteau reagent		

Table 2. Methods of estimating the total number of bacteria

Technique	Description
Counting chambers	The number of bacteria per ml of a suspension can be determined by means of a counting chamber. The best form of counting chamber is the Helber cell, but a hemocytometer is also useful. Suspensions of pathogenic or motile bacteria should be inactivated before they are counted, by heating at 55–60°C for 1 h. Care should be taken not to overheat the suspension, or agglutination may occur. Prepare a dilution which will give an average count of about 20 bacteria per small square. If necessary examine a drop of undiluted culture. A few drops of methylene blue may be added to the diluent to make the bacteria more easily seen, although too much stain may cause agglutination
	Place a No. 1 coverslip on the slide and then exactly fill the chamber (not the ditches) with the suspension using a fine pipette
	Find the grid under low power, and then use the ×40 objective for counting. Oil should not be used as it causes the coverslip to sag and consequently changes the volume of the liquid
	Count the number of organisms per small square, covering all planes of focus, in 20 squares taken diagonally across the grid. Repeat with a second mount. At least 1000 cells should be counted. The more cells counted, the greater the accuracy
	If a cell lies on the lines of the grid, only those on two sides of the square should be counted. All the component cells in a small clump of separate cells should be counted as single individuals
	From the number of organisms per square, the number per ml of the suspension is calculated. The volume of liquid lying over each square is given on each slide. In a Thoma hemocytometer, the depth of the chamber is 0.01 cm and the area of each square is 0.000025 cm^2. The number of bacteria per ml of the suspension is the average number per square × 4 × 10^6 × the dilution factor
	In the Helber cell the depth of the chamber is 0.002 cm and the area of each square is 0.000025 cm^2. Consequently, the number of cells per ml is the number per square × 2 × 10^7 × the dilution factor

Estimating Population Sizes of Bacteria

Continued

43

Table 2. Methods of estimating the total number of bacteria, *continued*

Technique	Description
The direct smear method	A known volume of bacteria in suspension is spread over a defined area of a microscope slide, stained and counted. A variety of different stains can be used (e.g. methylene blue) but fluorescein isothiocyanate (FITC), a specific protein stain, is very good. Bacteria stained with methylene blue appear blue when observed under white light and those stained with FITC fluoresce a green color under short-wavelength blue light
	Transfer 0.01 ml of the bacterial suspension to a 1 cm² area of a microscope slide. Allow the slide to dry at room temperature
Solutions needed	0.5 M carbonate–bicarbonate buffer (pH 9.6) 0.3 ml
	0.001 M phosphate-buffered saline (pH 7.1) 6.0 ml
	0.85% physiological saline 5.7 ml
	Crystalline FITC 5.3 mg
	Stain the slide in a damp chamber for 5 min. Wash gently but thoroughly, blot dry. Observe with a fluorescence microscope
Agar film technique	This technique has been used to count the number of organisms in a sample of soil
	The soil is sieved through a 2 mm sieve and a known weight of soil is placed in a small crucible with 5 ml of sterile distilled water and thoroughly ground up with a glass rod. The suspension is poured into a 100 ml sterile flask and the remainder washed into the same flask with a further 5 ml of water
	The soil suspension is made up to 50 ml with 1.5% agar (cooled). After shaking the flask vigorously leave for 3 sec to allow sedimentation of coarse grains. All this must be carried out at 45°C
	A sample is pipetted from just under the surface of the suspension on to the platform of a hemocytometer slide (0.1 mm depth), covered with a coverslip and left to set
	The hemocytometer slide is immersed in sterile distilled water and the coverslip removed. Surplus agar in the ditches should be removed with a scalpel. The hemocytometer is agitated and the loosened film floated on to an ordinary microscope slide and allowed to dry. Distilled water must be used to prevent crystallization of salts in tap water
	The films should be dried slowly at room temperature to prevent the agar splitting

The dry films are immersed for 1 h in the following stain:

Phenol (5% aqueous)   15 ml
Aniline blue (1% aqueous)   1 ml
Glacial acetic acid   4 ml
Filter 1 h after preparation

The stained films are washed, dehydrated in 95% ethanol
Twenty random fields are counted under a × 100 oil immersion lens on each of four replicate slides. Knowing the volume of agar observed in one microscope field (see hemocytometer instructions), the number of organisms in the original soil can be calculated

Table 3. Methods of counting the number of viable bacteria in a sample

Method	Description
Dilution plate/spread plate method	Tenfold dilutions are prepared. With the pipette used for mixing the dilution, 1 ml of this dilution is placed in each of three or more sterile petri dishes; then after making the next dilutions, 1 ml of this is treated in the same way and so on until the range of dilutions (normally three) is plated
	After mixing the dilutions, 15–20 ml of agar medium, previously melted and cooled to 45°C, are added to each plate and even distribution ensured (see Chapter 2)
	When set, the plates are incubated at the required temperature. After incubation, dilutions where the plates show more than 300 or fewer than 30 colonies are discarded. Errors due to overcrowding or to sampling make these counts unreliable. Thus only one dilution will actually be counted if tenfold dilutions are used
	Count all the colonies on each plate. Take the average of the counts for each plate and multiply by the dilution factor. The numbers should be expressed as the number of colony forming units (CFU) developing at x°C in y days on z medium, and not as the number of bacteria per ml or g of the original material

Continued

Estimating Population Sizes of Bacteria

45

Bacterial Cell Culture

Table 3. Methods of counting the number of viable bacteria in a sample, *continued*

Method	Description
	The assumption that one colony represents one bacterium is not justified. Bacteria may form clumps or chains which may or may not be broken during dilution. Actinomycete spores complicate the counting of these organisms as does the filamentous nature of this genera. Further, it is unlikely when counting a mixed flora that all the species present will develop equally well under the set incubation conditions of the count
	The inoculum may, as an alternative, be spread over the surface of prepared plates of agar. Usually an inoculum of 0.1 ml is then used and the agar is dried to remove surface water. The inoculum is spread with a sterile glass spreader
Roll tube method	This method is more economical with materials. Accurate counting is less easy, but the method is used to distinguish between gross and slight contamination in a large number of samples
	Dilutions are prepared and a large number of sterile tubes containing 2–3 ml of liquid agar medium (i.e. held at 45°C to prevent the agar setting) are inoculated with 1 ml of the dilution required as previously described. Replace each tube in the water bath at 45°C as soon as it has been inoculated
	When all the tubes have been inoculated, take the tubes from the water bath and roll them one by one under a stream of cold water so that the agar sets in a thin even film
	Incubate the tubes with the top downwards. Count the colonies after incubation and work out the results as described in the plate count method
Membrane filter method	This method is used for water samples where only a few organisms are present. It depends on the concentration of the inoculum rather than dilution
	Sterile cellulose acetate membranes (pore size approximately 0.45 µm) are placed in a sterile filter holder
	A known volume of the fluid is passed through the filter. This filter is then laid on a thin layer of medium in a petri dish and incubated. Bacteria held on the filter develop into colonies which can be counted. It is often difficult to count as an even distribution of colonies cannot be ensured. By using particular selective media, certain groups of bacteria can be selected

Most probable number (MPN) method

As well as being economical in terms of time and materials, this method can be used where the plate count cannot (e.g. with materials that contain species which form 'spreading' colonies on agar) or where it is desired to determine numbers of a particular group which is sparsely represented in material with a mixed bacterial flora

A series of tenfold dilutions are made so that when 1 ml samples of the highest dilution are inoculated into a suitable medium, very few, if any, bacteria are likely to grow. Thus, if a bacterial suspension contains 5×10^9 bacteria per ml, 1 ml of a 10^{-9} dilution would contain five organisms, and most or all the tubes of medium would show growth. In a 10^{-12} dilution, all five tubes of the medium would probably show no growth as there would only be one organism per 200 ml

Transfer three or more 1 ml replicate samples of each suitable dilution to tubes of liquid or plates of solid medium and incubate. A greater range of dilutions (usually five or six) are required for this method to enable accurate prediction of bacterial numbers

After incubation, record the presence or absence of growth in the cultures. Statistical tables have been compiled (e.g. McGrady tables) enabling the most probable number of organisms per ml of the suspension to be determined. An example of a McGrady table has been compiled for a dilution factor of 10, and for five replicate tubes inoculated from each dilution of culture (see Appendix)

The classic use of this method is in the 'presumptive coliform' count used in the examination of water supplies for fecal contamination. Other uses include the estimation of pathogenic fungi in flour

The entries in the McGrady table are the number of turbid tubes observed after inoculation of the same volume from any three successive tenfold dilutions. The stated MPN is the estimated mean viable count per inoculum taken from the most concentrated of the three dilutions being considered

Surface drop method (Miles and Misra)

This method is excellent for making a series of viable counts at short intervals on small samples of suspension containing a single bacterial species. It cannot be used for forms with 'spreading' or irregular colonies, or mixed suspensions

Dilutions are prepared by a drop method using a sterile pipette. The pipette must have a perfectly even tip and must be held vertically with even pressure so that a steady rate of dropping is achieved together with drops of equal volume

Place 45 drops of the diluent in each of six small sterile tubes. This will suffice for dilutions up to 10^{-6}. To the first tube add five drops of the bacterial suspension. Mix well and transfer five drops of the mixture to a second tube

Continued

Table 3. Methods of counting the number of viable bacteria in a sample, *continued*

Method	Description
	Ideally a fresh pipette should be used for each dilution. However, exactly equivalent pipettes are impossible to obtain and washing the pipette with distilled water may have to suffice
	If other dilutions are required (e.g. a twofold dilution series) then add 20 drops of bacterial suspension to 20 drops of diluent
	For plating out, two well dried agar plates are required. Mark the underside of each plate in six numbered sectors
	A sterile dropper pipette with an exactly known volume is needed for this step. Usually a pipette with a tip diameter of 0.94 mm, delivering a drop of 0.02 ml is used. Place a drop of the highest dilution in sectors one and four on each plate; then a drop of the next dilution in sectors two and five and a drop of the lowest dilution in sectors three and six. Each drop is therefore plated in quadruplicate. The drops should be absorbed quickly without spreading. After absorption, incubate the plates
	Count the number of colonies developed from each drop of each dilution. Take the mean count for each dilution and estimate the number of cells present in each ml of the original suspension. Ideally between 5 and 20 bacteria are ideal for each drop
Postgate's method (modified)	It is often useful to know what proportion of a culture is viable. This may be achieved using Postgate's method, though it must be pointed out that absolute numbers of viable cells are not counted, merely the proportion determined
	Molten agar is poured on to a sterile glass slide to a depth of approximately 1 mm
	Small 5 × 5 mm squares are cut from it with a sterile flat needle and transferred to a new slide
	Each square is inoculated with a dilution of the culture to be observed to give about 30 cells per microscope field
	After 5 min, the inoculated blocks are covered with a sterile coverslip and incubated aseptically in a damp chamber for a few hours until microcolonies of 8–16 cells are formed
	Dead cells will not divide and form microcolonies and so a count of single cells and colonies will allow the ratio of dead:living bacteria to be determined. Ten fields of view should be chosen at random on each of at least three replicate slides

Chapter 5   CHARACTERIZATION OF BACTERIA USING BIOCHEMICAL AND CULTURAL TESTS

1 Introduction

Because of the size of bacteria, morphological studies are never enough to show conclusively the delimitation of species. Other characters must therefore be used to allow more precise classifications of organisms to be made. These must include pathogenicity and serological tests, gene probing, DNA base composition and phage typing. Perhaps the most widely used tests, however, are those designed to demonstrate the biochemical and physiological characteristics of micro-organisms.

Previously, much emphasis has been placed on the importance of different characteristics in classifying bacteria. The characters selected as important for each organism are selected on a purely subjective basis, with different researchers giving different weight to any single character. An, as yet incomplete, reassessment of the classification of bacteria has taken place in which it has been asssumed that all characters are of equal importance until proved otherwise, and that groups of organisms are to be established on the basis of maximum correlation of characters. In the course of this assessment, many new biochemical and physiological tests have been employed and the following represents only a selection.

The biochemical methods used fall into four groups:

1. Mineral growth requirements and substrate utilization.
2. Detection of metabolic end-products.

Characterization Using Biochemical and Cultural Tests

3. Identification of characteristic compounds such as enzymes, polysaccharides and toxins.
4. Resistance and sensitivity to antibiotics and other chemicals.

2 Mineral growth requirements and substrate utilization

Knowledge of the ability of bacteria to dissimilate certain substrates and to synthesize various products is vital for the characterization of a bacteria. Routine qualitative tests designed to permit convenient detection of important biochemical features generally consist of a nutrient medium plus substrate in which the organism is cultured. *Table 1* describes some of the common tests used in bacteriology.

3 Detection of metabolic end-products

More definitive measurements of metabolic products are

5 Biochemical examination of anaerobic bacteria

For tests requiring plate cultures, anaerobic jars should be used for incubation, but by making slight modifications to the medium many of the tests can be carried out without this apparatus. The media may be rendered semi-solid by the addition of agar (0.5%). The precautions outlined previously (Chapter 3) must be observed. The iron nail method may be used but the formation of dark colored compounds may mask results.

Fermentative metabolism can be demonstrated by sealing cultures with Vaseline®.

6 Rapid testing methods

6.1 Multipoint inoculation

It is possible to carry out large numbers of some tests

often required. Such data enable one to establish better criteria for a taxonomic group. A range of tests commonly used to examine the metabolic activity of a range of bacteria are presented in *Table 2*.

4 Identification of characteristic compounds

Knowledge of the biochemical activities of a bacterial culture has many applications in biology beyond that of characterizing a species. The biochemical activities of bacteria make them either beneficial or harmful. *Table 3* describes a range of fermentation tests used in bacteriology.

Bacteria accomplish their various biochemical activities including growth and replication using nutrients present in culture medium. These biochemical transformations occur both inside and outside the bacterial cell and are mediated through the action of enzymes. The presence or absence of an enzyme can be used in the identification and characterization of bacteria. Some of the common tests used are presented in *Table 4*.

simultaneously by using a multipoint inoculator to inoculate petri dishes divided into several compartments.

6.2 API test strips

Analytical profile index (API) test strips, produced by Bio Merieux SA, may be used to obtain test results quickly. These consist of a series of miniature capsules on a molded plastic strip, each of which contains a sterile dehydrated medium in powder form. Addition of water containing a bacterial suspension simultaneously rehydrates and inoculates the medium. A rapid reaction is obtained because of the small volume of medium and the large inoculum used. The strip of microtubes is incubated for 18–24 h at 35–37°C. Reactions are evident from the color changes of indicator chemicals, either with or without addition of further reagents. These changes occur after incubation of the strip in a small humidified plastic chamber. The identification of the unknown bacterium is achieved by determining a seven digit profile index number and consulting the API profile recognition system.

The test strips can be used to identify specific groups of

Characterization Using Biochemical and Cultural Tests

organisms such as members of the Enterobacteriaceae, staphylococci, anaerobes, lactic acid bacteria, streptococci and clinical yeasts.

6.3 Enterotube II system

The basic philosophy of the Enterotube II system is the same as the API system. This system provides ease, speed and low cost in the identification of Gram-negative Enterobacteriaceae. The system consists of a single tube containing 12 compartments, each containing a different agar-solidified culture medium. Compartments that require aerobic conditions have small openings that allow in air. Those requiring anaerobic conditions have a layer of paraffin wax. There is a self-enclosed inoculating needle. This needle can touch an isolated bacterial colony and is then drawn through the 12 compartments, inoculating the test media. Fifteen standard tests are performed. After 18–24 h of incubation the color changes that occur in each of the compartments are recorded. Interpretation of the results is determined using a five digit code obtained from the results. A differential chart, supplied by the manufacturer, or a computed-based program, Enterobacteriaceae numerical coding and identification system for Enterotube (ENCISE) can be used.

7 Cultural characteristics

In addition to the morphological and biochemical characteristics, it is essential to record the appearance of bacterial colonies under defined conditions, for example on specific media such as agar streak plates (*Figure 1*), gelatin stabs or in-broth culture (*Figure 2*). Pigment production should also be recorded, preferably by comparing the color with a standard color chart such as those produced by Tresner, Backus & Prauser for streptomycetes or the Rembrandt color chart for bacteria. Pigments may be insoluble or they may diffuse into the medium and, in the case of streptomycetes, may be produced by different parts of the culture (i.e. aerial or submerged mycelium).

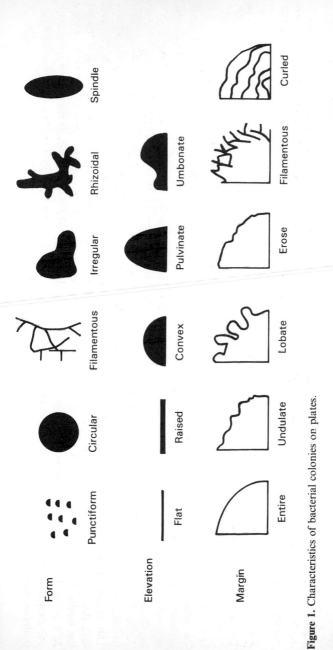

Figure 1. Characteristics of bacterial colonies on plates.

Characterization Using Biochemical and Cultural Tests

Bacterial Cell Culture

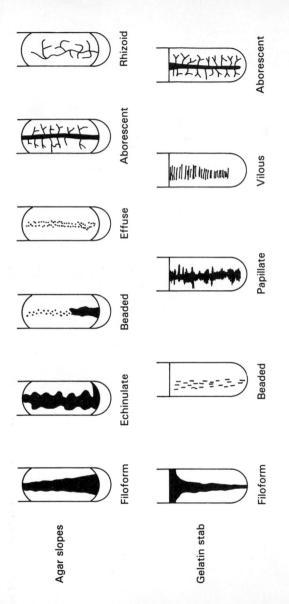

Agar slopes

Filiform   Echinulate   Beaded   Effuse   Aborescent   Rhizoid

Gelatin stab

Filiform   Beaded   Papillate   Vilous   Aborescent

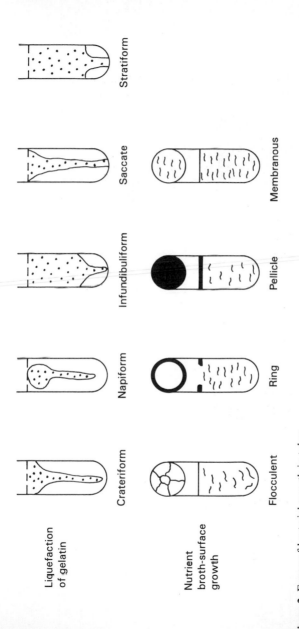

Figure 2. Forms of bacterial growth in tubes.

Characterization Using Biochemical and Cultural Tests

Bacterial Cell Culture

Table 1. Mineral growth requirments and substrate utilization tests used to examine bacteria

Type of test	Media		Description
Nutritional studies	NH₄H₂PO₄	1.0 g	It is often useful to know something about the nutritional requirements of a
	KCl	0.2 g	particular bacterium, from both a cultural and taxonomic standpoint
	MgSO₄.7H₂O	0.2 g	The growth of bacteria with an inorganic nitrogen source can be detected in
	Glucose	10.0 g	Ayer, Rupp and Johnson's medium. Several serial sub-cultures should be made
	Distilled water	1.0 l	before recording growth as positive since the inoculum may carry over organic
	Adjust pH to 7.0		nitrogen and growth factors. Further nutritional studies can be made by replacing
			sources of nitrogen in this medium for others
Environmental effects			The relationship between organisms and free oxygen and pH also falls into this group
			of tests. These relationships can be elucidated by methods outlined previously
			(e.g. agar shake cultures, the use of buffered media etc.)
Utilization of organic acids	KH₂PO₄	1.0 g	Some bacteria are able to utilize salts of certain organic acids as sole carbon sources.
	MgSO₄.7H₂O	0.2 g	Pathogenic species, which are nutritionally exacting, are less likely to be able to do
	Na(NH₄)HPO₄	1.5 g	this, whilst many of the less exacting bacteria, such as those commonly found in soils
	Distilled water	1.0 l	and water, can
	Sodium citrate	2.0 g	Koser's citrate medium is a mineral medium with sodium citrate as the sole carbon
			source
			Bromothymol blue may be added to detect slight growth by the alkaline reaction
			produced. Other organic acids can be used in place of citric acid (e.g. formic acid)

(Note on the subscripts as written in the table: $NH_4H_2PO_4$, $MgSO_4 \cdot 7H_2O$, KH_2PO_4, $Na(NH_4)HPO_4$.)

Table 2. Detection of metabolic end-products for the examination of bacteria

Type of test	Media		Description
Indole	p-Dimethylaminobenzaldehyde 95% Ethanol Ehrlich's reagent	2.0 g 100.0 ml 1.0 ml	Indole is produced by some bacteria from the amino acid tryptophan. A tube of peptone water is inoculated and samples of the culture are tested at intervals. This is then added to the culture and a maximum of 0.5 ml of concentrated HCl added drop by drop until a red zone appears between the alcohol and the peptone solution. If the colored solution is soluble in amyl alcohol or chloroform it can be considered to be indole
Hydrogen sulfide			Hydrogen sulfide may be produced from sulfur-containing amino acids, such as hydrogen cysteine, present in peptone. A test for hydrogen sulfide production can be combined with the above test. When the peptone water is inoculated, a filter paper strip impregnated with a saturated lead acetate solution is inserted between the plug and the tube, taking care to keep it dry. If hydrogen sulfide is produced, the paper will become blackened Some bacteria also reduce inorganic sulfur compounds. A chemically defined medium containing a sulfur compound (e.g. sulfate or thiosulfate) can be used instead of peptone water. Only organisms that can utilize inorganic nitrogen can be tested in this way A number of special media containing iron salts and a source of sulfate can also be used (e.g. Kliger's iron agar). Hydrogen sulfide production is recognized by blackening of the medium
Acetylmethyl carbinol (Vosges–Proskauer test)	Glucose Peptone K₂HPO₄	0.5 g 0.5 g 0.5 g	Acetoin or acetylmethyl carbinol can be produced from glucose via pyruvic acid. Two enzymes are required for this reaction, pyruvate and acetolactic decarboxylase. Acetoin may be reduced to 2,3-butanediol

Continued

Characterization Using Biochemical and Cultural Tests

57

Bacterial Cell Culture

Table 2. Detection of metabolic end-products for the examination of bacteria, *continued*

Type of test	Media		Description
			and so give a negative result. Therefore in sensitive tests 100 ml of distilled water should be used to detect acetoin
			The normal method of detection is to oxidize acetoin to diacetyl which reacts with guanidine residues in the medium to give a cherry red color. To prevent reducing conditions from developing, fumarate is sometimes added as a hydrogen acceptor or else glucose is replaced by pyruvate
			After 3 days of incubation, a 2 ml sample of culture fluid is tested by adding 5 ml of 40% KOH and a trace of creatine, shaking vigorously and allowing to stand for up to 60 min. The culture is positive if a cherry red color develops
Methyl red test	Methyl red	0.4 g	In examination of coliform bacteria, the VP test is carried out at the same time as the methyl red test. The latter determines the hydrogen ion concentration produced in a defined time and specific medium. The glucose–phosphate medium described above must be used and not modified. Test the culture after 3 days at 30°C
	Distilled water	100.0 ml	To one part of the culture add the methyl red solution (1 drop ml^{-1} culture). A magenta red color is positive, a yellow color is negative
Reduction of nitrates			Some organisms are capable of reducing nitrate to nitrite or even to ammonia
			The test organisms are grown in peptone water containing 1% potassium nitrate. After incubation, the culture is divided into three portions. One portion is tested by adding a few drops of sulfanilic

acid reagent, followed by a few drops of α-naphthylamine reagent. A red color indicates the presence of nitrite. The production of free nitrogen may be detected by allowing it to collect in a Durham tube

A second portion of culture is tested with Nessler's reagent for the presence of ammonia, but since many organisms produce ammonia when grown in peptone broth, care must be taken when interpreting results. A sample of uninoculated medium should also be tested since some peptones give a positive reaction (i.e. an orange–brown precipitate) with Nessler's reagent

If no nitrite is detected, the third portion of the culture should be tested for residual nitrate, as it is possible that all the nitrate will have been converted firstly to nitrite and then to nitrogen or ammonium. To do this add a small quantity of zinc dust. This will chemically reduce residual nitrate to nitrite, and this can be tested for as above. If the presence of nitrite is proved it follows that the bacteria did not reduce nitrite to nitrate; rather it was the zinc dust which reduced it. If the test on the third portion is negative for nitrite, it follows that all the nitrate was reduced to another product via nitrite by the bacteria, which was thus a positive nitrate-reducing bacterium

Characterization Using Biochemical and Cultural Tests

Table 3. A range of fermentation tests used to enumerate bacteria

Fermentation test	Media		Description
Hugh and Liefson's method	Peptone	0.2 g	Certain carbohydrates (i.e. sugars, alcohols and glucosides) produce quantities of acid and/or gas when attacked. It is important to know which acid or gas it is. Generally simple tests for detecting the nature of the acids are lacking and so results are recorded merely as to whether gas and/or acid was produced. One of the most convenient tests for this purpose is the Hugh and Liefson's test
	Sodium chloride	0.5 g	
	K_2HPO_4	0.03 g	
	Bromothymol blue	0.3 g	
	Carbohydrate	1.0 g	
	Agar	0.3 g	
	Distilled water	100.0 ml	A semi-solid agar medium is put in a tube to a depth of 4 cm. Two tubes are required for each culture and sugar tested. The bromothymol blue is dissolved in water and 0.3 ml of a 1% solution added to each 100 ml of medium.
	pH 7.1		

Alcoholic solutions of indicators should not be used, as acid may be produced from the alcohol. For critical studies, the carbohydrate should be sterilized separately and added to the otherwise complete sterile medium

After inoculation, one of the pairs of tubes is covered by a layer of sterile melted Vaseline® to a depth of 1 cm

Several types of reaction may be observed:

Fermentative bacteria which phosphorylate glucose and then split it into two triose molecules before forming acid will produce an acid reaction in both tubes (or the sealed tube only)

Oxidative bacteria which convert the aldehyde group to a carboxyl group in glucose to form glucuronic acid will produce an acidic reaction in the open tube only, leaving the sealed tube unchanged. The acid reaction produced by oxidative organisms is apparent first at the surface, and then extends

Production of aesculin

gradually downwards into the medium. Organisms which oxidize glucose but do not ferment it have never been observed to ferment any other carbohydrates. Therefore the sealed tube can be omitted when testing the other sugars

Hydrolysis of aesculin is shown by the production of blackening in the aesculin medium

Peptone	10.0 g
Aesculin	1.0 g
Ferric citrate	0.5 g
Distilled water	1.0 l
Agar	1.2 g

Levan production

Many bacteria including plant pathogens are characterized by the ability to produce levan polysaccharides from sucrose. Such compounds may also be produced by soil bacteria. A plate of nutrient agar containing 5% sucrose is poured, and a channel cut in the agar. This channel is filled with the same medium with 0.1% aniline blue added to it. Polysaccharide-producing organisms attract the dye towards them, whilst others remain colorless. The inoculum can be added in two ways: as streaks to the left of the channel, and as individually spaced colonies to the right of the channel

Characterization Using Biochemical and Cultural Tests

Bacterial Cell Culture

Table 4. Identification of enzymes used to characterize bacteria

Enzyme	Assay	Description
Diastase		The breakdown of starch is more complex than that of sugars because hydrolysis must occur before it can be used by bacteria. It is possible to substitute starch for a sugar in one of the above fermentation reactions, but this gives little information about the extent of the reaction. A more reliable method is to prepare a plate of nutrient agar containing 0.2% soluble starch. This is inoculated by streaking the test bacteria across the diameter of the plate. The plate is incubated for 2–7 days and then flooded with Lugol's iodine. The breadth of the clear zone outside the area indicates the extent of starch degradation. A second plate containing a similar medium, with bromocresol purple as indicator, may be used to detect acid production
		Another method for organisms which are able to utilize inorganic nitrogen is to incorporate 0.001% of soluble starch in a mineral base medium. Samples are tested for starch at intervals with iodine. This method is particularly sensitive
Cellulase and chitinase		As far as is known cellulase and chitinase are enzymes possessed by comparatively few species of bacteria. Decomposition is relatively slow. The simplest method for testing for activity is to inoculate a tube of peptone water in which a strip of filter paper or chitin is partly submerged. If a mineral medium is used instead of peptone water, a heavy inoculum should be used. This is done because a bacterium might not produce sufficient enzyme to be detectable unless growth has started
		'Halo' methods can also be used. Finely divided cellulose or chitin can be incorporated in agar, and the zones of clearing noted. Usually the cellulose or chitin has to be ball-milled to give particles of about 1 μm diameter, otherwise sedimentation occurs and the bacterium cannot get near the substrate. Alternatively, cellulose and chitin

Pectinase

can be dissolved in concentrated hydrochloric acid, re-precipitated by pouring into water, and then incorporated into agar in a colloidal suspension

The ability to decompose the middle lamella of plant cell walls is an important property of certain plant pathogens, especially when first isolated. The ability to liquefy a calcium pectate gel may be tested, but some difficulty may be encountered since several enzymes are actually responsible for decomposing the various pectic substances

It is often more useful to test samples of culture fluid (after inactivating or removing the bacteria) for the ability to destroy thin discs of plant tissue immersed in them

Similar techniques can be developed for detecting enzyme action of other complex carbohydrates (e.g. xylan, agar etc.)

Lipases

The ability to decompose fats may be a useful taxonomic character and is of interest to food microbiologists. The addition of 1% tributyrin (the triglycerol ester of butyric acid) to nutrient agar produces a slight opaque medium on which lipolytic enzymes produce clear zones. Care must be taken in interpreting results since butyric acid is liberated during decomposition. This may be toxic to certain organisms leading to false negative results

Olive oil, butter and other fats can be added to a basal medium of any type and solidified with agar. Five per cent fat is added to the molten agar and the medium shaken well. It may be preferable to sterilize the fat separately (1 h at 150°C in the hot air oven). Immediately before use the medium must be shaken vigorously to emulsify the fat. After incubation, plates should be examined under a low-power microscope. The fat globules under and around lipolytic colonies appear granular. The addition of dyes such as Nile blue is often recommended

Characterization Using Biochemical and Cultural Tests

Continued

63

Enzymes in Molecular Biology

Table 4. Identification of enzymes used to characterize bacteria, *continued*

Enzyme	Assay		Description
Urease	Peptone	1.0 g	The ability to decompose urea is also a useful diagnostic character in routine work, for instance in the identification of certain pathogens in feces. Thus *Salmonella* and *Shigella* are urease negative while nonpathogenic *Pseudomonas* and *Proteus* are urease-positive
	NaCl	5.0 g	
	KH$_2$PO$_4$	2.0 g	
	Phenol red	0.012 g	The basal medium may be a chemically defined one with a simple carbohydrate as a source of energy and urea as a nitrogen source, or it may contain peptone (e.g. Christensen's medium). The development of a strong alkaline reaction (red coloration) after growth indicates the breakdown of urea and the formation of ammonia. Some bacteria can produce ammonia from peptone and acid from sugars. Christensen's medium is balanced in order to minimize these errors. Controls should be set up if there is any doubt. A 20% solution of urea, sterilized by filtration, is added aseptically to a sterile, cool basal medium to give a final concentration of 2%. Allow the medium to set as a slope
	Glucose	1.0 g	
	Agar	20.0 g	
	Distilled water	1.0 l	
	pH 6.8–6.9		
Proteolytic activity			The ability to attack proteins may be used as a taxonomic criterion and may have great ecological significance. A range of different proteins can be used to demonstrate the action of bacteria
	Gelatin		Liquefaction or hydrolysis of gelatin is investigated using the 'deep stab' method which involves inoculating a tube of nutrient gelatin with the test organism. If the cultures are incubated at 20°C, the area of liquefaction may take a characteristic form, but this may be variable, possibly relating to the oxygen requirement of the culture. Cultures showing liquefaction should be cooled under a tap since some gelatin samples melt spontaneously at 24°C

Smith's modification of the Frazier plate is a useful technique for detecting weak liquefiers. A culture is streaked on to a plate of nutrient agar containing 0.4% gelatin. This is incubated at 25°C for 2–4 days. The plate is covered with 8–10 ml of a solution of 15 g $HgCl_2$ in 100 ml of distilled water and 20 ml of HCl (12 M). A white opaque precipitate is formed where the gelatin has not been liquefied

A third method can be used with liquid culture (i.e. peptone water). A small charcoal gelatin disc is added to the broth in a test tube and the culture inoculated. Gelatin hydrolysis is shown when charcoal is liberated into the medium. The method is a convenient way of counting gelatin liquefying organisms using the most probable method

Caseolytic activity

This is usually tested by inoculating a plate of milk agar (5% sterile milk added to melted and cooled water agar, well mixed and poured immediately). The medium is opaque, and clear zones of hydrolysis appear around caseolytic colonies. Casein hydrolysis can be detected by the Frazier plate method described above for gelatin

Litmus milk

Skimmed milk with sufficient litmus to give a lilac color is often used, especially in dairy bacteriology. Bromocresol purple may be substituted for litmus. Changes of reaction are noted together with the formation of sweet or stormy clots. The latter of these clots indicates the production of gas, presumably by the fermentation of lactose, and clearing which denotes casein digestion. Clots may be formed by acid precipitation of casein or by the action of bacterial enzymes (rennin)

Blood serum

Serum is a complex mixture of proteins and other constituents. The ability to liquefy heat coagulated serum is of diagnostic importance in some groups. The serum, sterilized by filtration and distributed into sterile tubes, is heated to 72°C until it has just solidified, the tube being held in a sloping position. It should not be overheated. Inoculation is performed in the normal way

Characterization Using Biochemical and Cultural Tests

Continued

Enzymes in Molecular Biology

Table 4. Identification of enzymes used to characterize bacteria, *continued*

Enzyme	Assay		Description
Deaminases	Yeast extract	3.0 g	Some bacteria deaminate phenylalanine forming phenylpyruvate by hydrolysis with phenylalanine deaminase. To test for this grow the bacteria on the medium for at least 24 h at 37°C. Flood the culture with 4–5 drops of ferric chloride solution (10% w/v). A green color in the agar indicates the presence of phenylpyruvate
	D,L-Phenylalanine	2.0 g	
	Na_2HPO_4	1.0 g	
	NaCl	5.0 g	
	Distilled water	1.0 l	
	Agar	15.0 g	
Hemolytic activity			Many bacteria, not necessarily pathogens, produce substances which cause complete or partial lysis of the red blood cells of various animals. Fresh blood agar is used, for example nutrient agar to which 5–10% of sterile (aseptically collected) citrated or defibrinated blood (usually horse blood) is added. Half the contents of a 25 ml bottle of nutrient agar is poured into a sterile dish, the remainder being kept free from contamination. When the first layer has set, add 1 ml of blood to the cooled agar in the bottle which should still be liquid. Mix well and pour over the agar to form a thin layer. The nutrient agar used should be slightly more concentrated to allow for the dilution by the blood
			β-Hemolytic streptococci produce complete hydrolysis. α-Hemolytic strains cause partial hemolysis and produce methemoglobin. In the former the medium is clear, while in the latter the colonies are surrounded by a greenish zone. The latter are generally less pathogenic
Oxidase			The oxidase and cytochrome oxidase tests are essentially identical reactions. The addition of α-naphthol to the oxidase test if done rapidly will convert it to a cytochrome oxidase test. The oxidase test is performed by putting 2–3 drops of fresh 1% aqueous tetramethylparaphenylene-diamine (TMPD) dihydrochloride on to a

piece of paper in a line 3–6 mm long. The colony turns purple in 5–10 sec if oxidase is present

An alternative method is to pour a small quantity of very freshly prepared 1% TMPD solution on to a slope culture. The formation of a deep purple color within 10 sec indicates the presence of an oxidase

Decarboxylases

Amino acid decarboxylases are tested for using Moeller's basal medium to which is added either 1% L-arginine monohydrochloride or 1% L-ornithine hydrochloride. If D, L-acids are used, 2% solutions are needed. Inoculate together with a control

Add a 4.5 mm layer of sterile paraffin oil to all tubes. Examine after 4 days. Positive reactions are shown by a color change from yellow to violet or reddish-violet

Peptone	5.0 g
Beef extract	5.0 g
Bromocresol purple (1.6%)	0.625 ml
Cresol red (0.2%)	2.5 ml
Glucose	0.5 ml
Pyridoxal	5.0 mg
Distilled water	1.0 l
pH 6.0	

Autoclave 4 ml amounts for 10 min

Catalase

Catalase decomposes hydrogen peroxide to produce oxygen and water. It is present in most aerobic organisms. Its function is to remove H_2O_2 which forms as a result of coupled oxidation–reduction processes involving oxygen

Pour 1 ml hydrogen peroxide (10 vol.%) over the surface of a 1 day agar slope culture. Release of oxygen bubbles indicates the presence of catalase

Characterization Using Biochemical and Cultural Tests

Chapter 6 PCR

1 Introduction

The polymerase chain reaction is an *in vitro* nucleic acid amplification method. The reaction is based around the repeated thermal cycling of the reaction mixture thereby dissociating the products of the previous thermal cycle, then allowing the association of these dissociated products. During the reaction substrate DNA is denatured at high temperature (e.g. 94°C) to give single-stranded target molecules. Short oligonucleotide primers, amplimers, then anneal to a specific nucleotide sequence on the target molecule at lower temperatures (30–60°C). The thermal cycle is concluded by the amplimers being extended enzymatically through the addition of appropriate based-paired deoxynucleotidetriphosphates (dNTPs) at an intermediate temperature (e.g. 72°C), producing another double-stranded copy of the original target. If each cycle of PCR was 100% efficient, each cycle would result in the

with thermal circulators are important factors for PCR. Thermal cyclers vary in their levels of sophistication and care must be taken when purchasing the cycler to ensure that the specifications match your requirements.

3 Extraction of DNA

There are a number of published methods for the extraction of DNA and RNA. Most of these methods are based on the affinity purification of nucleic acids. This method usually eliminates the requirement for solvent extraction using chemicals such as phenol and reducing the number of centrifugation steps. For most PCR applications, the DNA or RNA extracted can be diluted prior to PCR. This enables rather crude nucleic acid preparations to be used success-fully. However, where the number of target nucleic acid doubling of the number of copies of the original target sequence. The PCR process therefore enables trace amounts of a target molecule to be detected from a complex background. PCR may be used to analyze specific DNA targets, or after a reverse transcription step, RNA sequences can also be amplified as DNA copies. The applications of PCR to bacterial culture include the identification of bacteria in clinical samples and phylogenetic comparative analysis to determine evolutionary relationships between bacteria.

2 Requirements for use of PCR

Contamination of samples for amplification poses a major problem to PCR. A range of physical and chemical techniques can be used to reduce this. Physical separation of the three different areas involved in PCR namely, DNA extraction, PCR preparation and PCR analysis, reduces the chances of contamination from accidental PCR product carry-over. A number of companies manufacture both the thermocycler and the consumables required for PCR (see Chapter 8 for a list of supplier's names and addresses). The parameters associated

sequences is low (e.g. in DNA extracted from environmental samples or in testing for pathogenic bacteria in food samples) and dilutions cannot therefore be made, the nucleic acid extracted must be purer. Examples of the crude preparation for the isolation of bacterial DNA is presented in *Table 1*. A large range of kits now exist for the extraction and purification of nucleic acids. The names and suppliers can be found in Chapter 8.

4 DNA polymerases

Polymerases catalyze the synthesis of polynucleotide chains from monomeric dNTPs using one of the parental strands as a template for the synthesis of the complementary strand. DNA polymerases require a short segment to anneal to a complementary sequence to prime synthesis which proceeds in a 5' to 3' direction. Thermostable DNA polymerases can be added at the beginning of the amplification without further additions. A range of commercially available thermostable polymerase enzymes are available from the suppliers listed in Chapter 8.

5 Primers

Primers are designed to maximize both the specificity and efficiency amplification of the PCR reaction. Specificity and efficiency are determined by primer parameters such as position, base composition, length and melting temperature. Primers used in PCR are generally 18–30 nucleotides in length. If possible primers should be made so that the GC content is within 40–60%. Primer pairs should also be designed so that there is no complementarity of the 3' ends either inter or intra individual primers. Chemically labeled PCR primers can be readily produced using automated solid-phase oligonucleotide synthesis. Examples of chemicals which can be added include biotin, digoxigenin, dinitrophenyl and fluorescent dyes such as fluorescein.

6 Optimization of PCR

For routine screening purposes in which suitable conditions have already been established it is not necessary to enhance the production of a specific PCR product. For preparative PCR however, it may be necessary to enhance a specific PCR product. PCR optimization kits are available from a range of suppliers. However, individual parameters of the PCR protocol can be optimized to improve specific PCR production. *Table 2* summarizes parameters which may be altered to improve PCR.

7 Detection of PCR amplified products

Not all PCRs yield a single product; nonspecific priming may generate products from which the target of interest may be difficult to detect. A variety of methods used detect PCR products are presented in *Table 3*.

8 Purification of PCR products

If the main aim of the PCR was not to detect a product it may be necessary to purify the target product. Generally the PCR product will be checked via gel electrophoresis. Subsequent purification of an excised gel band is necessary. *Table 4* illustrates the different techniques available for purification of PCR products.

Table 1. Examples of protocols for the isolation of crude preparations of DNA for PCR

Sample	Method
Bacterial liquid culture	1. Remove a 200 µl aliquot of bacterial culture into a microcentrifuge tube
	2. Centrifuge in a microcentrifuge at full speed for 2 min. Discard the supernatant
	3. Resuspend the pellet in 200 µl sterile distilled water
	4. Place the microcentrifuge in a boiling water bath for 5 min
	5. Centrifuge for 5 min at full speed
	6. Use between 10 and 30 µl of the supernatant per PCR
Bacterial plate colony	1. Pick a bacterial colony and suspend in 200 µl water in a microcentrifuge tube
	2. Continue from step 2 above
	3. Use between 1 and 10 µl of the final supernatant per PCR
Soil and sediments	1. Aliquot 1 g samples of sediment into two 2 ml screw-cap Eppendorf tubes, each containing 0.5 g of 0.1 mm diameter glass beads baked at 260°C
	2. Add the following to each tube: 0.7 ml of 120 mM sodium phosphate buffer (pH 8) containing 1% (w/v) acid-washed polyvinylpolypyrrolidone, 0.5 ml of Tris-equilibrated phenol (pH 8) and 50 µl of 20% (w/v) sodium dodecylsulfate
	3. Bead beat the samples three times at 2000 rev min^{-1} for 30 sec, with 30 sec on ice between bead beatings
	4. Centrifuge at 12000 g for 2 min. Store the supernatants on ice
	5. Resuspend the pellet in 0.7 ml of 120 mM phosphate buffer
	6. Bead beat at 2000 rev min^{-1} for 30 sec and then re-centrifuge
	7. Combine the supernatants from both the first and second extractions
	8. Make an HTP (bio-Gel HTP; Bio-Rad Laboratories Ltd, Hemel Hempstead, UK) spin column from a 1 ml plastic syringe plugged with a sterile glass wool plug supporting 0.6 ml HTP
	9. Load the extracted samples on to the spin column by spinning 0.7 ml aliquots of extract at 100 g in a swing-out rotor at room temperature for 3 min until all the sample is loaded

Continued

PCR

Table 1. Examples of protocols for the isolation of crude preparations of DNA for PCR, *continued*

Sample	Method
	10. Wash the column by spinning three times with 0.5 ml aliquots of 120 mM phosphate buffer (pH 7.2) to remove protein binding to the HTP
	11. Elute nucleic acid into a sterile Eppendorf tube with 0.4 ml of 300 mM dipotassium phosphate (pH 7.2)
	12. Desalt the eluent with a 2.5 ml Sephadex G-75 spin column
	13. Precipitate the nucleic acid with 2.5 volumes of ethanol
	14. If the pellet produced is white, resuspend in 50 µl of TE buffer (10 mM Tris, 1 mM EDTA, (pH 8)
	15. If the pellet is brown, indicating some humic contamination then resuspend the pellet in 200 µl of sterile distilled water and precipitate with polyethylene glycol 8000 at 4°C for 1 h
	16. After centrifugation at 12000 *g* for 15 min removed the supernatant by aspiration, wash the pellet with 70% ethanol and finally resuspend in 50 µl of TE buffer

Table 2. Key PCR parameters which can be adjusted to improve target production

Parameter	Description
Reaction components	*Template*: the ideal template for a PCR is free from contaminants such as nucleases. Amplification may be affected by the base composition of the template
	Buffer composition: optimum buffer composition varies according to the enzyme and PCR application. Tris-HCl, Tris-acetate, or Tricine may be used as a buffering agent at pH values from 8.3 to 8.8. Mg^{2+} concentration can exert a major effect on the efficiency of PCR due to its complexing with dNTPs and the Mg^{2+} requirement of the enzyme. Buffers may also be supplemented with auxiliary components known as PCR enhancers
	Polymerase: the choice and concentration of polymerase will affect PCR optimization. Optimum enzyme conditions lie between 0.005 and 0.025 units μl^{-1}
	Primers: the selection of primers has an important influence on the efficiency and specificity of the PCR amplification. Purity, base composition and length of primers will all influence the generation of amplification products. Concentration of primers added to a reaction will also influence yield, and it may be necessary to titrate the final concentration in the range 0.1–1.0 μM
Thermal cycling	When optimizing PCR complete thermal equilibrium of the reaction mix must be achieved. Therefore reaction volumes and tube wall thickness are important factors to consider when setting up cycle parameters. Most reactions are performed in 25–100 μl volumes in 200–500 μl glass capillary tubes
	Denaturation: it is important that the first denaturation step of a PCR is completed so that the DNA template strands are fully separated. When using complex genomic DNA, denaturation may have to occur at 100°C for 3–5 min; however, generally 94–96°C for 2–3 min is sufficient. Following this initial denaturation a temperature of 92–95°C will be adequate for the remainder of the PCR
	Annealing: calculating the temperature at which primers anneal specifically to a template is of fundamental importance to the efficiency of amplification. However, the melting temperature (T_m) of any given primer can only act as an approximate guide. Increasing the annealing temperature above the T_m generally increases specificity but reduces yield. Decreasing the temperature reduces the specificity of amplification

Continued

PCR

Table 2. Key PCR parameters which can be adjusted to improve target production, *continued*

Parameter	Description
	Polymerization: Extension rates of polymerases are generally between 2 and 4 kbp min^{-1}. Complete strand synthesis of up to 5 kbp should be complete within 3 min. For longer targets, extension steps may be increased up to 20 min. Generally increasing extension times will enhance yield
PCR	*Nested PCR*: specificity can directly influence product yield and length. Amplimers of 20–30 bp usually provide sufficient specificity for PCR. However, where the target is present in low copy number, or absolute specificity is required, nested PCR is the most efficient method. Nested primers annealing internally to the first primers are used in a secondary PCR to amplify primary reaction products, thereby producing a smaller amplification product. The increase in specificity produced by using nested PCR is due to the fact that the chances of nontarget sequences being amplified is much reduced
	Long PCR: PCR usually involves the generation of products less than 5 kb in length. Occasionally larger products (10–30 kb) may be required. Consideration of the reaction conditions must be made

Table 3. Methods used for the detection of PCR products

Method	Description
Homogeneous detection (single step detection)	In this method the presence or absence of amplified sequences, although not the specificity of the PCR, is determined. Each PCR reaction is set up with an internal detection system such as ethidium bromide, a dye that intercalates between double-stranded DNA helices which then fluoresce when illuminated with UV light. The ingredients in a PCR are single stranded and the products are double stranded; therefore the incorporation of ethidium bromide in a PCR results in an exponential increase in fluorescence with each amplification cycle. The specificity of the reaction is unaffected by the presence of ethidium bromide. Ethidium bromide can be added to the reaction mix (4 µg ml^{-1}) and fluorescence detected at either 254 or 312 nm using a UV transilluminator
Gel detection of products	Size separation through agarose gel electrophoresis is the usual method for the detection of PCR products. Samples are loaded on to gels and an electric current is applied. Separation and subsequent visualization of reaction products enables analysis of products and estimation of yields. For identification of bands, elution and re-amplification followed by sequencing offers one of the best methods. A number of steps are necessary for the preparation and running of gels:

Sample preparation: to a sample of PCR product add an equal volume of loading buffer (to a final concentration of × 1). Loading buffers contain dye and increase sample density. To make a sixfold concentration prepare a solution containing 0.35% bromophenol blue and 40% (w/v) sucrose in distilled water

Preparation of agarose gels: electrophoresis of PCR products through an agarose gel is the usual method of determining PCR products. Migration of DNA fragments through agarose is dependent on the agarose concentration and the molecular weight of the DNA fragment along with buffer composition. Use DNA-grade agarose (0.8–1.2 % w/v) for the casting and running of the gel. To make a tenfold stock solution of TAE add 48.4 g of Tris base and 11.4 ml glacial acetic acid to 20 ml 0.5 M EDTA (pH 8.0). Boil the agarose solution until the agarose has melted. Ethidium bromide (0.5 µg ml^{-1}) can be added to the agarose for staining. Pour the agarose into a gel cast. The comb can be inserted before or after pouring the gel. Make sure the agar is not too hot (see the manufacturer's guidelines) |

Continued

Table 3. Methods used for the detection of PCR products, *continued*

Method	Description
	Running the gel: flood the gel tank with TAE buffer. Clean each well by flushing with buffer. Pipette the samples into each well and begin electrophoresis. Ensure that the gel temperature remains constant throughout the gel. Measure the distance between the two electrodes and apply a voltage of between 1 and 8 V cm^{-1}. *Product visualization:* ethidium bromide is the most common method of detecting products. Upon examination of the gel under UV light, products will be observed as bright bands
Membrane detection	*Membrane:* immobilization of DNA on to a solid support followed by hybridization to an internal probe enhances the sensitivity of specific product detection when compared to gel electrophoresis. Two classes of solid support can be used for blotting. Nylon binds DNA irreversibly while nitrocellulose binds DNA through reversible hydrophobic interactions. Most of these membranes are produced with a 0.45 or 0.20 µm filter *Probes:* probes for the detection of specific DNA products may be obtained commercially. Radionucleotides may also be used

Table 4 Methods available for the purification of targeted PCR products

Method	Description
1. DNA isolation kits	A variety of commercial kits are available for the isolation of DNA from agarose (see Chapter 8). These kits may be used to purify PCR products directly from the reaction mix
2. Ultrafiltration spin cartridges	This procedure relies on the repeated dilution and concentration of the reaction mix. During the concentration step, low molecular weight compounds pass through the membrane filter
3. Magnetic separations	This uses streptavidin-coated magnetic beads (MagneSphere®, Promega) where a biotinylated amplimer has been employed or a biotin dNTP incorporated
4. HPLC	High performance liquid chromatography may also be used to purify PCR products

Chapter 7 ISOLATION AND CULTURE OF ACTINOMYCETES

1 Introduction

Actinomycetes are an important member of the soil-microbial community, and can be isolated from all soils. Rich garden soil or alkaline soils will contain high numbers of actinomycetes. Although actinomycetes are closely related to bacteria, they have some features in common with fungi. As a result the methods applied to them are something of a mixture of those applied to fungi and bacteria. Morphological features are used to identify genera but biochemical and physiological tests are used for species identification. In addition the wide range of pigments produced by this group are often used in their taxonomy.

2 Microscopic examination of actinomycetes

The dimensions of actinomycetes are similar to those of bacteria and so observations with light microsocopy are mainly made under high-power and oil-immersion objectives. Stains similar to those used for bacteria are sometimes used, but often the delicate sporing apparatus is disrupted by staining. Observations are therefore often made directly without staining.

A convenient way to observe many actinomycetes is to grow them on coverslips inserted at a 45° angle into a suitable solid medium. The coverslip is carefully removed from the medium, placed on a slide and then observed. The use of an electron microscope for examination of actinomycetes is important and many additional morphological features, which are not visible under the light microscope, can be seen.

A light microscope fitted with a high-power long-working

distance objective allows the direct examination of plate colonies, without damaging spore structures. With experience, this procedure permits the rapid screening for representatives of individual taxa.

3 Culture of actinomycetes on artificial media

Many media have been used to grow actinomycetes and some are used for specific purposes (i.e. to stimulate antibiotic production or to stimulate pigment production). The majority of actinomycetes prefer a neutral to alkaline pH (pH 8.0) which should be maintained in isolation media. The media listed in *Table 1* are often used for general cultivation.

When transferring actinomycetes from one culture to another, the same procedure as used for bacteria is applied. They are transferred with a loop and streaked over the surface of the medium. Although mycelial-like fungi, they have a much more restricted spread over media, and large colonies

5 Preliminary identification of actinomycetes

Once isolated the diverse physiological nature of actinomycetes can be used for preliminary identification before a more detailed biochemical analysis. *Table 3* lists a number of key microscopic observations which can be used to identify actinomycetes to the species level.

6 Identification of the *Streptomyces* species

The genus *Streptomyces* contains many antibiotic-producing strains and has therefore received much attention from industrial bacteriologists. Over 400 species have now been described and many criteria are used to identify them. The most common criteria used for identification include:

1. Morphology of spore chains: straight, flexous, looped, spiral, verticillate.
2. Ornamentation of spores as seen under the electron microscope: smooth, spiny, warty, hairy.
3. Color characteristics of spores and pigments produced in

are only obtained if the inoculum is spread over the surface of the medium.

In liquid culture actinomycetes behave rather like fungi, forming a surface pellicle in static culture, or pellets in agitated culture. A few genera (e.g. *Amycolata*) can lose their mycelial form and break into rod-like elements which grow in liquid culture in a similar way to bacteria.

4 Isolation of actinomycetes from natural sources

Although actinomycetes are mainly isolated from soils and composts, they can readily be isolated from aquatic sediments. A variety of media has been used for their selective isolation and a number are shown in *Table 2*.

culture medium. A wide range of colors are produced by *Streptomyces* and attempts have been made to standardize descriptions using color codes.

4. Physiological–biochemical tests: production of black melanin pigment from peptone or tyrosine, use of different carbon sources.

7 Antibiotic production

Many well known antibiotics are produced by *Streptomyces* species, for example neomycin by *S. fradiae*, streptomycin from *S. griseus*. Antibiotics may be active against bacteria fungi, protozoa or viruses. Only a few have been chemically synthesized and most are produced by growing the streptomycete in optimal conditions for production. The production of an antibiotic (or antibiotics) by a streptomycete can be demonstrated as shown in *Table 4*.

Isolation and Culture of Actinomycetes

Bacterial Cell Culture

Table 1. Media for the cultivation of actinomycetes

Medium	Preparation		Comments
Oatmeal agar	Oatmeal	56.0 g	Oatmeal is cooked and filtered. The filtrate is made up to 1 l with water and agar added
	Agar	12.0 g	
	pH8		
Starch–nitrate agar	Starch	10.0 g	
	KNO_3	3.0 g	
	NaCl	2.0 g	
	K_2HPO_4	1.0 g	
	$MgSO_4.7H_2O$	0.05 g	
	$CaCO_3$	0.02 g	
	$FeSO_4.7H_2O$	0.05 g	
	Agar	12.0 g	
	Distilled water	1.0 l	

Some genera (e.g. *Amycolata*) grow well on nutrient agar which is used for bacteria

Table 2. Isolation of actinomycetes from different sources

Source	Media		Comments
Terrestrial forms	Starch	10.0 g	Many actinomycete genera occur in soil and most of the antibiotic-producing strains of *Streptomyces* originate from soil. They may be isolated using the dilution technique, but it is necessary to try and prevent the growth of bacteria and fungi on the isolation plates. This may partly be achieved by using a medium with constituents chosen to favor the growth of actinomycetes rather than fungi or bacteria. One such medium is starch-casein
	Casein	0.3 g	
	KNO$_3$	2.0 g	
	NaCl	2.0 g	
	K$_2$HPO$_4$	2.0 g	
	MgSO$_4$.7H$_2$O	0.05 g	
	CaCO$_3$	0.02 g	The selectivity of such a medium can be further improved by adding antibiotics to suppress growth of fungi and other bacteria such as actidione (50 μg ml^{-1})
	FeSO$_4$.7H$_2$O	0.01 g	
	Agar	12.0 g	
	Distilled water pH 8	1.0 l	
Thermophilic actinomycetes			Some actinomycetes (e.g. *Thermomonospora*) only develop at temperatures above 40°C in culture. These organisms occur in heated materials such as wet hay and compost heaps. They can be isolated by using rich organic media (e.g oatmeal agar) and incubating plates at high temperatures
Aquatic forms			Some genera (e.g. *Streptosporangium*) occur in fresh water. These may be isolated by baiting water with a suitable substrate such as potato pieces. Pollen grains have also been successfully used as baits for aquatic actinomycetes
Anaerobic actinomycetes			A few genera (e.g. *Actinomyces*) are anaerobic and can be detected and grown using the procedures described for anaerobic bacteria

Isolation and Culture of Actinomycetes

Table 3. Key for the preliminary identification of actinomycetes

Step	Test	Species	Proceed to step No.
1	Substrate mycelium fragmenting into bacteria-like elements	–	If yes go to 2 If no go to 3
2	Grows aerobically Grows anaerobically	*Amycolata* *Actinomyces*	Yes Yes
3	No aerial mycelium Aerial mycelium	*Micromonospora* –	Yes If yes go to 4
4	Spores on aerial and substrate mycelium Spores only on aerial mycelium	– 	If yes go to 5 If yes go to 6
5	Spores appear in short chains Spores appear as single spores Spores appear in clusters	*Micropolyspora* *Thermoactinomycetes* *Thermomonospora*	Yes Yes Yes
6	Spores appear in pairs Spores present in long chains or spirals Spores are single but appear densely packed Spores found in sporangia	*Microbispora* *Streptomyces* *Saccharomonospora* *Streptosporangium*	Yes Yes Yes Yes

Table 4. Procedure for the identification of antibiotic activity by actinomycetes

Step	Protocol
1	Take a plate culture of a streptomycete and cut out 12 cylinders from the colony with a sterile glass tube
2	Seed three vials of molten medium (45°C) with test organisms (e.g. Gram-positive bacterium, Gram-negative bacterium and a fungus)
3	Pour the seeded media into dishes and leave to set
4	Place four cylinders from the streptomycete colony on to the surface of each seeded plate and incubate at 25°C
5	After a few days, any antibiotics produced by the streptomycete will have diffused out of the cylinder into the seeded medium and if it is effective against a test organism, a clear zone of inhibition will be visible

Isolation and Culture of Actinomycetes

Chapter 8 MANUFACTURERS AND SUPPLIERS

The major suppliers of reagents, cultures and equipment described in this book are listed below. This list is by no means exhaustive since new products are continually being produced. Both the US and the UK addresses of companies are given when appropriate. Names and addresses of local companies can be obtained from the addresses given. Telephone and fax numbers are given.

1 Main sources of bacterial cultures

North America
American Type Culture Collection (ATCC), 12301 Parklawn Drive, Rockville, MD 20852-1776, USA.
Tel (301) 881 2600, (800) 638 6597.
Fax (301) 231 5826.

National Collections of Industrial and Marine Bacteria Ltd, 23 Saint Machar Drive, Aberdeen AB24 3RY, UK.
Tel (01224) 273 332.
Fax (01224) 487 658.

2 Suppliers of staining solutions and reagents

Aldrich Chemical Company Co. Inc., 1001 West Saint Paul Avenue, Milwaukee, WI 53233, USA.
Tel (900) 558 9160. (414) 273 3850.
Fax (800) 962 9591, (414) 273 4979.

Baxter Healthcare Corp., Scientific Products Division, 8855 McGaw Road, Columbia, MD 21045, USA.
Tel (800) 234 8401, (301) 290 8418.

Carolina Biological Supply Company, 2700 York Road, Burlington, NC 27215, USA.
Tel (910) 584 0381, (800) 334 5551.
Fax (910) 584 3399.

Connecticut Valley Biological Supply Company Inc., 82 Valley Road, (PO Box 326), Southampton, MA 01073, USA.
Tel (413) 527 4030, (800) 628 7748.
Fax (800) 355 6813.

Ward's Natural Science Establishment, Inc., PO Box 1712, Rochester, NY 14603, USA.
Tel (716) 3589 2502.
Fax (716) 334 6164.

Europe

National Collection of Type Cultures (NCTC), Central Public Health Laboratory, 61 Colindale Avenue, London NW9 5HT.
Tel (0181) 200 4400.
Fax (0181) 205 7483.

Fisher Scientific Company, Corporate Headquarters, 2000 Park Lane, Pittsburgh, PA 15275, USA.
Tel (412) 490 8300.

ICN Biomedicals, Inc., 3300 Hyland Avenue, Costa Mesa, CA 92626, USA.
Tel (714) 545 0100.
Fax (714) 557 4872.

Sigma Chemical Company, PO Box 14508, Saint Louis, MO 63178, USA.
Tel (800) 325 3010, (314) 771 5750.
Fax (800) 325 5052, (314) 771 5757.

Sigma Chemical Company Ltd., Fancy Road, Poole, Dorset BH17 7NH, UK.
Tel (01202) 733 114, (0800) 447 788.
Fax (01202) 715 460.

Manufacturers and Suppliers

3 Sources of media and supplies

Becton Dickinson Microbiology Systems, PO Box 243, 250 Schilling Circle, Cockeysville, MD 21030, USA.

Carolina Biological Supply Company, 2700 York Road, Burlington, NC 27214, USA.
Tel (919) 584 0381, (800) 334 5551.
Fax (919) 584 3399.

Connecticut Valley Biological Supply Company Inc., PO Box 326, Southampton, MA 01073, USA.
Tel (413) 527 4030, (800) 628 7748.
Fax (800) 355 6813.

Difco Laboratories, PO Box 331058, Detroit, MI 48232, USA.
Tel (313) 462 8500.
Fax (313) 462 8517.

Oxoid USA Inc., 9017 Red Branch Road, Columbia, MD 21045, USA.

Scott Laboratories, Fiskville, RI 02823, USA.

Anachem Ltd, Anachem House, 20 Charles Street, Luton, Bedfordshire LU2 OEB, UK.
Tel (01582) 456 666.
Fax (01582) 391 768.

BDH Laboratory Supplies, Merck Ltd, Hunter Boulevard, Magna Park, Lutterworth, Leicestershire LE17 4XN, UK.
Tel (0800) 223 344.
Fax (0145) 555 8586.

Bio-Rad Laboratories, Main Office, 2000 Alfred Noble Drive, Hercules, CA 94647, USA.
Tel (510) 741 1000, (800) 424 6723.
Fax (510) 741 1060, (800) 879 2289.

Bio-Rad Laboratories Ltd, Bio-Rad House, Maylands Avenue, Hemel Hempstead, Hertfordshire HP2 7TD, UK.
Tel (01442) 232 552, (0800) 181 134.
Fax (01442) 259 118.

4 Suppliers of reagents and equipment for use in PCR analysis

Amersham North America, 2636 South Clearbrook Drive, Arlington Heights, IL 60005, USA.
Tel (847) 593 6300, (708) 593 6300.
Fax (847) 437 1640/70.

Amersham International plc, Amersham Place, Little Chalfont, Bucks HP7 9NA, UK.
Tel (01494) 544 000, (0800) 515 313.
Fax (01494) 542 266, 0800 616 927.

Amicon, 72 Cherry Hill Drive, Danvers, MA 01923, USA.

Amicon, Upper Mill, Stonehouse, Gloucester GL10 2BJ, UK.
Tel (01453) 825 183.
Fax (01453) 826 686.

Boehringer Mannheim USA, 9115 Hague Road, PO Box 50414, Indianapolis, IN 46250-0414, USA.
Tel (0800) 262 1640.
Fax (317) 576 2754.

Boehringer Mannheim UK, Bell Lane, Lewes, East Sussex BN17 1LG, UK.
Tel (01273) 480 444, (0800) 521 578.
Fax (01273) 480 266, (0800) 181 087.

CLONTECH Laboratories Inc., 1020 East Meadow Circle, Palo Alto, CA 94303-4607, USA.
Tel (415) 424 8222, (800) 662 2566.
Fax (415) 424 1064, (800) 424 1350.

CLONTECH, Cambridge Biosciences, Stourbridge Common Business Centre, Swanns Road, Cambridge CB5 8LA, UK.
Tel (01223) 316 855.
Fax (01223) 607 32.

Manufacturers and Suppliers

Eppendorf-Netheler-Hinz GmbH, Biotech Products, Barkausenweg 1, Hamburg D-22339, Germany.
Tel 40 53801 0.
Fax 40 53801 593.

Fisons Scientific Equipment, Bishop Meadow Road, Loughborough, Leicestershire LE11 ORG, UK.
Tel (01509) 231 166.
Fax (01509) 231 893.

Flowgen Instruments Ltd, Shenstone, nr Lichfield, Staffordshire WS14 0EE, UK.
Tel (01543) 483 054.
Fax (01543) 483 055.

Gilson Inc., Box 27, 3000 West Beltline Highway, Middleton WI 53562-0027, USA.
Tel (608) 836 1551.
Fax (608) 831 4451.

Jencons Scientific Ltd, Cherrycourt Way Industrial Estate, Stanbridge Road, Leighton Buzzard, Bedfordshire LU7 8UA, UK.
Tel (01525) 372 010.
Fax (01525) 379 547.

Life Sciences International Ltd, Chadwick Road, Astmoor, Runcorn, Cheshire WA7 1PR, UK.
Tel (01928) 566 611.
Fax (01928) 565 845

Millipore Corporation, PO Box 255, Bedford, MA 01730, USA.
Tel (617) 275 9200.
Fax (800) 225 1380.

Millipore Corporation, The Boulevard, Blackmore Lane, Watford, Hertfordshire WD1 8YW, UK.
Tel (01923) 816 375.
Fax (01923) 818 297.

Life Technologies Inc., 8400 Helgerman Court, PO Box 6009, Gaithersberg, MD 20884 9980, USA.

Life Technologies Ltd, 3 Fountain Drive, Inchinnan Business Park, Paisley, Renfrewshire PA4 9RF, UK.
Tel (0141) 814 6100.
Fax (0141) 814 6287.

Grant Instruments (Cambridge) Ltd, Barrington, Cambridge CB2 5QZ, UK.
Tel (01763) 260 811.
Fax (01763) 262 410.

Hoefer Scientific Instruments, 654 Minnesota Street, PO Box 77387, San Francisco, CA 94107, USA.
Tel (415) 282 2307, (800) 227 4750.
Fax (415) 821 1081.

International Biotechnologies Inc., PO Box 9558, New Haven, CT 06535, USA.
Tel (203) 786 5600.

New England Biolabs Inc., 32 Tozer Road, Beverly, MA 01915-5599, USA.
Tel (508) 927 5054, (800) 632 5227.
Fax (508) 921 1350.

New England Biolabs (UK) Ltd, 67 Knowl Piece, Wilbury Way, Hitchin, Hertfordshire SG4 0TY, UK.
Tel (01462) 420 616, (0800) 318 486.
Fax (01462) 421 057.

Pharmacia Biotech Inc., 800 Centennial Avenue, PO Box 1327, Piscataway, NJ 08855-1327, USA.
Tel (800) 526 3593
Fax (800) 329 3593.

Pharmacia Biotech Ltd, 23 Grosvenor Road, St Albans, Hertfordshire AL1 3AW, UK.
Tel (01727) 814 000, (0800) 318 353.
Fax (0800) 318 354.

Manufacturers and Suppliers

Sarstedt Ltd, 1025 St James's Church Road, PO Box 468, Newton, NC 28658, USA.
Tel (704) 465 4000.
Fax (704) 465 4003.

Sarstedt Ltd, 68 Boston Road, Beaumont Leys, Leicester LE4 1AW, UK.
Tel (01533) 359 023.
Fax (01533) 366 099.

Scientific Imaging Systems Ltd, 36 Clifton Road, Cambridge CB1 4ZR, UK.
Tel (01223) 242 813.
Fax (01223) 243 036.

Sigma Chemical Company, 3050 Spruce Street, St Louis, MO 63103, USA.
Tel (314) 771 5750, (800) 325 3010.
Fax (314) 771 5757, (800) 325 5052.

United States Biochemical Corporation, PO Box 22400, Cleveland, OH 44122, USA.
Tel (216) 765 5000, (800) 321 9322.
Fax (216) 464 5075, (800) 535 0998.

United States Biochemical, Amersham Life Science, Amersham Place, Little Chalfont, Buckinghamshire HP7 9NA, UK.
Tel (01494) 544 000, (0800) 515 313.
Fax (01494) 542 266.

Whatman International Ltd, Whatman House, St Leonards Road, Maidstone, Kent ME16 0LS, UK.
Tel (01622) 676 670.
Fax (01622) 677 011.

5 Useful manuals for media preparation

DIFCO Manual (1984), *Dehydrated Culture Media and Reagents for Microbiology*, 10th Edn, Difco Laboratories, PO Box 331058, Detroit, MI 48232, USA.

Sigma-Aldrich Chemical Company Ltd, Fancy Road, Poole, Dorset BH12 4HQ, UK.
Tel (01202) 733 114, (0800) 373 731.
Fax (01202) 715 460, (0800) 378 785.

Power, D.A. and McCuen, P.J. (1988) *Manual of BBL Products and Laboratory Procedures*, 6th Edn, Becton Dickinson Microbiology Systems, PO Box 243, 250 Schilling Circle, Cockeysville, MD 21030, USA.

Manufacturers and Suppliers

FURTHER READING

Chapter 1

Atlas, R.M. and Parks, L.C. (eds) (1983) *Handbook of Microbiological Media*. CRC Press, Boca Raton, FL.

Gerhardt, P. *et al.* (eds) (1981) *Manual of Methods for General Microbiology*. American Society for Microbiology, Washington, DC.

Isenberg, H.D. (ed) (1992) *Clinical Microbiology Procedures Handbook*. American Society for Microbiology, Washington, DC.

Vesley, D. and Laner, J. (1986) in *Laboratory Safety: Principles and Practices* (B.M. Miller, ed.), p. 182. American Society for Microbiology, Washington, DC.

Chapter 2

Block, S.S. (ed.) (1991) *Disinfection, Sterilization, and Preservation*, 4th Edn. Lea & Febiger, Philadelphia.

Issac, S. and Jennings, D. (1995) *Microbial Culture*, Introduction to Biotechniques. BIOS Scientific Publishers, Oxford.

Balows, A. *et al.* (1992). *The Prokaryotes: A Handbook on the Biology of Bacteria: Ecophysiology, Isolation, Identification, Applications*, 2nd Edn. Springer-Verlag, New York.

Holt, J.G. *et al.* (eds) (1994) *Bergey's Manual of Determinitive Bacteriology*, 9th Edn. Williams & Wilkins, Baltimore, MD.

Chapter 5

Balows, A. *et al.* (eds) *Manual of Clinical Microbiology*, 5th Edn. American Society for Microbiology, Washington, DC.

Priest, F. and Austin, B. (1993) *Modern Bacterial Taxonomy*, 2nd Edn. Chapman & Hall, London.

Skerman, V.B.D. *et al.* (1989) *Approved Lists of Bacterial Names* (amended edition). American Society for Microbiology, Washington, DC.

Perkins, J.J. (1983) *Principles and Methods of Sterilization in Health Sciences*, 2nd Edn. Charles C. Thomas, Springfield, IL.

Chapter 3

Boatman, E.S. *et al.* (1987) *Bioscience* **37**, 384.

Gerhardt, P. *et al.* (eds) (1994) *Methods for General Bacteriology*. American Society for Microbiology, Washington, DC.

Rawlings, D.J. (1992) *Light Microscopy*. BIOS Scientific Publishers, Oxford.

Chapter 4

Meynell, G.G. and Meynell, E. (1970) *Theory and Practice in Experimental Bacteriology*. Cambridge University Press, Cambridge.

Koch, A.L. (1981) in *Manual of Methods for General Bacteriology* (P. Gerhardt, ed.). American Society for Microbiology, Washington, DC.

Chapter 6

Ellingboe, J. and Gyllensten, U.B. (eds) (1992) *The PCR Technique: DNA Sequencing*. Eaton Publishing, Natick, MA.

Erlich, H.A. (ed.) (1989) *PCR Technology*. Stockton Press, New York.

Griffin, H.G. and Griffin, A.M. (eds) (1994). *PCR Technology: Current Innovations*. CRC Press, Boca Raton, FL.

Innis, M.A. *et al.* (eds) (1990) *PCR Protocols: A Guide in Methods and Applications*. Academic Press, San Diego, CA.

McPherson, M.J. *et al.* (eds) (1991). *PCR: a Practical Approach*. Oxford University Press, Oxford.

McPherson, M.J. *et al.* (eds) (1995). *PCR-II: a Practical Approach*. Oxford University Press, Oxford.

Mullis, K.B. *et al.* (eds) (1994) *The Polymerase Chain Reaction*. Birkhauser, Boston.

Newton, C.R. and Graham, A. (1994) *PCR*. BIOS Scientific Publishers, Oxford.

Further Reading

APPENDIX

Values of the MPN for five tubes inoculated from each of three successive tenfold dilutions

Significant number — Numbers of replicate tubes showing turbidity at three successive dilutions			Most probable number	Significant number — Numbers of replicate tubes showing turbidity at three successive dilutions			Most probable number
0	0	1	0.18	5	0	1	3.1
1	0	0	0.20	5	1	0	3.3
1	1	0	0.40	5	1	1	4.6
2	0	0	0.45	5	2	0	4.9
2	0	1	0.68	5	2	1	7.0
2	1	0	0.68	5	2	2	9.5
2	2	0	0.93	5	3	0	11.0
3	0	0	0.78	5	3	1	11.0
3	0	1	1.1	5	3	2	14.0
3	1	0	1.1	5	4	0	13.0
3	2	0	1.4	5	4	1	17.0
4	0	0	1.3	5	4	2	22.0
4	0	1	1.7	5	4	3	28.0
4	1	0	1.7	5	5	0	24.0

4	1	1	2.1	5	5	1	35.0
4	2	0	2.2	5	5	2	54.0
4	2	1	2.6	5	5	3	92.0
4	3	0	2.7	5	5	4	160.0
5	0	0	2.3	5			

INDEX

THE ESSENTIAL DATA SERIES

Vectors P. Gacesa & D.P. Ramji
0 471 94841 1 1994 £14.99/$24.95

Human Cytogenetics
D.E. Rooney & B. H. Czepulkowski (Eds)
0 471 95076 9 1994 £14.99/$24.95

Animal Cells: Culture and Media
D.C. Darling & S.J. Morgan
0 471 94300 2 1994 £14.99/$24.95

Light Microscopy C.P. Rubbi
0 471 94270 7 1994 £14.99/$24.95

Gel Electrophoresis D. Patel
0 471 94306 1 1994 £14.99/$24.95

Centrifugation
D. Rickwood, T.C. Ford & J. Steensgaard
0 471 94271 5 1994 £14.99/$24.95

Available in 1997...

Cytokines A. Meager & C.J. Robinson (Eds)
0 471 97294 0 due October 1997 £14.99/$24.95

Growth Factors A. Meager & C.J. Robinson (Eds)
0 471 97295 9 due October 1997 £14.99/$24.95

The above two books are also available as a set:

Growth Factors/Cytokines
0 471 97296 7 due October 1997 £27.50/$44.00

Enzyme Assays S.K. Sreedharan & K. Brocklehurst
0 471 96527 8 due November 1997 £14.99/$24.95

ORDER FORM

Please send me:

Qty	Title	Price/copy	Total

All prices are correct at time of going press but subject to change.

Your order will be processed without delay, please allow 21 days for delivery.

We will refund your payment without question if you return any unwanted book to us in a re-saleable condition within 30 days.

All books are available from your bookseller.

Method of payment

☐ Payment £/$ _____ enclosed (Payable to John Wiley & Sons Ltd).

Orders for one book only - please add £2.00/$5.00 to cover postage and handling. Two or more books postage FREE.

☐ Purchase order enclosed ☐ Please send me an invoice (£2.00/$5.00 will be added to cover postage and handling).

☐ Please charge my credit card account

☐ American Express ☐ Diners Club ☐ Visa ☐ Mastercard

Card no: _____ Expiry date: _____

Signature: _____

Telephone our Customer Services Dept with your cash or credit card order on (01243) 843206 or dial FREE on 0800 243407 (UK only)

Send my order to:

Name (PLEASE PRINT)

Position:

Address:

_____

_____

_____

Telephone

Signature: _____ Date: _____

Return to:

Andrea Sharp, John Wiley & Sons Ltd, Baffins Lane, Chichester, West Sussex, PO19 1UD, UK

Fax: +44 1243 770460

or: Wiley Liss, 605 Third Avenue, New York, NY 10158-0012, USA

Fax: (212) 850 8888

☐ If you do not wish to receive mailings from other companies please tick this box or notify the Marketing Services Dept at John Wiley & Sons

WILEY